THE

CAMBRIDGE COURSE

OF

ELEMENTARY PHYSICS.

PART FIRST.

COHESION, ADHESION, CHEMICAL AFFINITY, AND ELECTRICITY.

BY W. J. ROLFE AND J. A. GILLET,
TEACHERS IN THE HIGH SCHOOL, CAMBRIDGE, MASS.

BOSTON:
CROSBY AND AINSWORTH.
NEW YORK: O. S. FELT.
1868.

University Press: Welch, Bigelow, & Co.,
Cambridge.

PREFACE.

An attempt is made in this Course to present the leading principles of Physics which are established at the present time, in a suitable form for teaching in our High Schools and Academies. No attempt has been made to write text-books "for schools and colleges," since the authors believe that such books are suitable for neither the one nor the other.

The authors have not sought to make these books encyclopædias of facts, as is too often the case with text-books on Physics, especially on Chemistry. They have sought to develop leading principles in the simplest and clearest manner possible; and as all the principles of Physics rest on facts determined by observation, they have adopted the plan of first giving the experiments which establish the facts, and of then drawing out the principle.

The experiments given are, in almost every case, those which one of the authors has used in the class-room. The aim has always been to give the simplest experiment (and the one requiring the simplest apparatus) that will establish the required fact.

Some two years ago, one of the authors was appointed teacher of Physics in the Cambridge High School. He at first tried text-books, but soon found that there was no book suitable for High School use, which would enable the scholar to gain any satisfactory knowledge of the present state of the physical sciences. The books served only to confuse the scholars, and to give them a distaste for the study of even so interesting a subject as Chemistry. He then determined to throw aside text-books, and to rely wholly on oral instruction. The scholars soon became interested, and made much more rapid progress.

But there are obvious objections to instruction wholly oral, especially with young scholars. Hence it becomes desirable to embody this instruction in a book, which the authors trust other teachers may find serviceable.

These books are not, however, designed to do away with oral instruction. In our school the lessons, whenever it is possible, are illustrated by experiments before the scholar is called upon to recite them; and the success which has attended this method may justify us in recommending it to other teachers.

Part II. of the Course will be devoted to Sound, Light, and Heat; Part III. to Gravity and Astronomy. There will probably be a small supplementary volume on elementary Mechanics. Part III. is nearly ready for the press, and will appear next.

Each Part will be complete in itself, and, of course, they can be used in any order; but the order given is believed to be the simplest and best.

Under AFFINITY the aim has been, first, to develop the great laws of combination, and then to illustrate the action of affinity in the processes of combustion, respiration, and decay, the growth of plants, destructive distillation, and the manufacture of the alkalies.

In this Course an atom is regarded, according to the present theory, as a portion of matter *chemically* indivisible. The new atomic weights, which, as is well known, are often double the old ones, are used; and sufficient prominence is given to the doctrine of *atomicity*, or the *atom-fixing* power of the elementary atoms. The authors have followed Miller, who is regarded as the best English authority on chemistry, in using both the unitary and dualistic symbols of the ternary salts, according to convenience; also, in using barred symbols for those elements whose atomic weights are now regarded as double the old equivalent weights.

In ELECTRICITY, greater prominence is given to voltaic and magneto-electricity than to frictional electricity, since, in the present state of the science, and of its applications in the arts, they are far more important. In frictional electricity, especial care has been taken to develop Faraday's beautiful theory of induction, which binds into one harmonious whole the otherwise heterogeneous mass of electrical phenomena. For this reason frictional electricity is presented last. Electricity is uniformly treated as a force, since the hypothesis of electric fluids is clearly out of date.

The most important applications of the electric force are pointed out. The principles of telegraphing, electrotyping, plating, and gilding are fully explained.

The decimal system of weights and measures will be mainly used in this Course.

The Appendix to Part I. contains full directions for performing all experiments that are at all difficult.

Free use has been made of all the material at our command. Many of the woodcuts have been copied from the pages of standard French and English works; the rest have been made from our own drawings. In preparing the sections on COHESION and ADHESION we have made constant use of two most excellent treatises on Chemical Physics; one by Prof. J. P. Cooke of Harvard College, and the other by Dr. W. A. Miller, of King's College, London. No teacher can afford to be without these works. The books most consulted in the preparation of the CHEMICAL AFFINITY are Hofmann's "Modern Chemistry" (London, 1865), Part Second of Miller's "Elements of Chemistry" (3d edition, London, 1864), and Roscoe's "Elementary Chemistry" (London, 1866). Great assistance has also been derived from a variety of other sources. In addition to the works mentioned above, we would call the attention of teachers to Faraday's "Lectures on the Chemical History of a Candle," on "The Physical Forces," (both reprinted by the Harpers, of New York,) and on "The Non-Metallic Elements"; and Professor Cooke's Lectures on "Religion and Chemistry" (Scribner, New York). These are all most excellent books. The material of the ELECTRICITY has been mainly drawn from Faraday's "Researches in Electricity," Dr. Noad's "Manual of Electricity," and a very valuable elementary work on Electricity prepared for Chambers's Educational Course by Dr. Ferguson, and recently published (1866) at Edinburgh.

The authors are jointly responsible for everything in the book. The plan of the Course has been mainly developed by Mr. Gillet, who has also selected, arranged, and partially elaborated the material of this Part; while Mr. Rolfe has given the material its final elaboration, and superintended the passage of the book through the press. There has been mutual consultation at every point. There will be the same division of labor in the preparation of the forthcoming Parts.

CAMBRIDGE, May 1, 1867.

TABLE OF CONTENTS.

I.

COHESION.

COHESION.

1. *Matter is made up of Molecules.* — When a piece of ice is heated to a temperature of 32° it melts and becomes water. The parts of the ice hold together firmly, while those of the water into which it is converted move among themselves with the greatest ease. When ice melts, then, it is evidently resolved into minute particles, which retain but a slight hold upon one another. Wax, resin, lead, iron, gold, and many other substances, also melt when they are heated to a certain degree of temperature. Most solids, then, by means of heat, can be resolved into minute particles, which move freely among themselves.

If water be heated to a certain degree of temperature, it boils and becomes steam. Its particles are still further separated from one another. What is true of water in this respect is found by experiment to be true of other liquids.

The particles into which a solid is resolved when it melts, and a liquid when it boils, are called *molecules.* Molecule means *a little mass,* and of these little masses all matter is supposed to be made up.

2. *Molecules are exceedingly small.* — It is impossible to pulverize a solid so finely as to convert it into a liquid. A piece of gold may be divided into particles so small that each can barely be made out with the most powerful microscope, yet the gold is solid still. When heated, however, the pulverized gold is converted into a liquid; that

is, each minute piece is resolved into particles which move freely among themselves. Hence these molecules are much too small to be seen with the best microscope.

3. *The Molecules are not in actual contact.* — If a brass ball, which at the ordinary temperature will just pass through a ring, be plunged into a freezing mixture and left until it becomes very cold, it will then pass through the ring very easily, not touching it at all. What is true of a brass ball in this respect is found to be true of every solid.

Fig. 1.

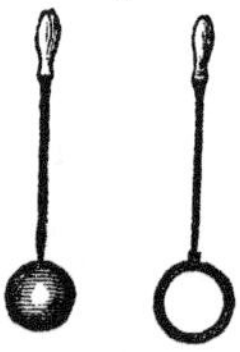

If a bulb with a projecting tube be filled with water up to a certain point on the tube, and the bulb be then plunged into a freezing mixture, the water will fall in the tube; and the same is found to be true if any other liquid be put into the bulb.

Fig. 2.

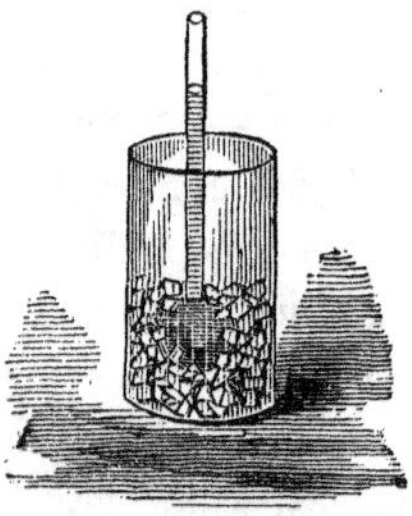

If a similar bulb be filled with air, and the end of the tube be held under water, and the bulb be cooled by means of a freezing mixture, the water at once rises in the tube; showing that the air occupies less space when cooled. The same is found by experiment to be true of other gases.

Fig. 3.

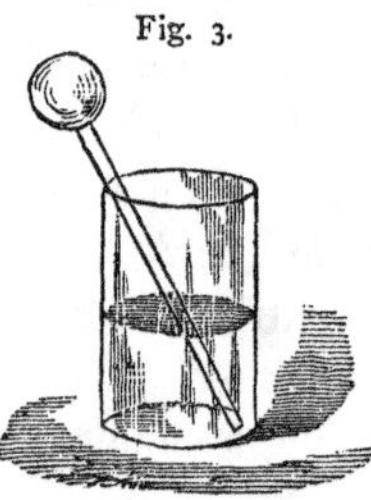

We find, then, that solids, liquids, and gases contract when cooled; and there seems to be no limit to this contraction, for they continue to contract, however much they are cooled.

Now, when a body contracts, its molecules are supposed to come nearer together, and since, so far as we know, a body may continue to contract indefinitely, it follows that the molecules are never in actual contact.

4. *The Spaces between the Molecules are immense in comparison with the Size of the Molecules.* — Though the spaces between the molecules are very minute, since they cannot be discerned even with the most powerful microscope, there are good reasons for believing that they are immense when compared with the molecules themselves.

The molecules of a body have been compared to the earth, sun, moon, and stars, and the spaces between the molecules to the spaces between these heavenly bodies. This comparison is probably very near the truth. If we imagine a being small enough to live on one of the molecules in the centre of a stone, as we live on the earth, such a being, on looking out into the space about him, would see here and there, at immense distances, other molecules, as we see the scattered stars in the heavens at night. The molecules, though exceedingly minute, are perfectly distinct and definite masses, like the earth, moon, and stars, and they are separated by spaces many thousand times as great as that occupied by each molecule.

5. *An Attractive and a Repulsive Molecular Force.*—If we attempt to pull any solid asunder, we perceive at once that the particles of which it is composed are held together more or less firmly. That which holds them together is called an *attractive force.* If a glass rod be dipped into water, a drop hangs from its end when taken out. This drop is made up of molecules which are evidently held together. In the case of liquids the molecules are held together but feebly, and the attractive force seems to be slight.

If a rubber bag partially filled with air, and closed so as to be air-tight, be placed under the receiver of an air-pump, and the air exhausted from the receiver, the air within the bag will at once expand, as is shown by the filling out of the bag. The same is found to be true when the bag is partially filled with any other gas; showing that gases when left to themselves expand, that is, their molecules separate.

The force which separates the molecules is called a *repulsive force.*

Since these forces act between molecules, they are called *molecular forces.*

6. *These two Forces act together.* — A brass ball (see Fig. 1) which will just pass through a ring at the ordinary temperature, will not pass through the ring after being heated; showing that the ball expands when heated. By similar experiments it is found that all solids expand when heated. While a solid is heating, then, a repulsive force must be acting, which separates the molecules. If, however, while the solid is heating, we attempt to pull it asunder, it resists; showing that the molecules are still held together by an attractive force.

We see by the foregoing experiment that heat is the repulsive force which separates the molecules. As the temperature of the solid rises this repulsive force grows stronger and stronger, until it nearly equals the attractive force, when the solid melts, that is, becomes a liquid.

If a glass bulb with a projecting tube (see Fig. 2) be partially filled with any liquid and then heated, the liquid rises in the tube; showing that the liquid expands, and that the repulsive force increases. When the repulsive force exceeds the attractive, the liquid boils, that is, it is converted into a gas.

From these facts we conclude that the attractive and repulsive forces are always acting together, and that the different conditions of matter depend upon their comparative strength.

7. *The Three States of Matter.*—When the attractive force is considerably stronger than the repulsive force, matter is in *the solid state;* when the two forces are nearly balanced, in *the liquid state;* and when the repulsive force is the stronger, in *the gaseous state.*

8. *Cohesion and Adhesion.* — The force which holds the

molecules of a solid or liquid together is evidently the *excess* of the attractive over the repulsive force; for if the two forces were just equal, they would just neutralize each other, and the molecules would not be held together in the least.

In the case of iron or water, it is evident that molecules of *the same kind* are held together. When, however, we mark on a blackboard with a piece of chalk, or write on paper with ink, it is equally evident that molecules of *different* kinds are held together.

The force which holds together molecules of the same kind is called *cohesion*; that which holds together molecules of different kinds, *adhesion*.

9. *These Forces act only through insensible Distances.*—Two hemispheres of lead will not cohere unless perfectly smooth and clean, and pressed firmly together so as to seem to be in actual contact, when they cohere quite strongly. Plates of glass, from simply resting upon each other in the warehouse, have been known to cohere so firmly that they would break elsewhere as readily as where they came in contact.

10. *Solids.* — Matter, as we have seen, exists in three states, the *solid*, the *liquid*, and the *gaseous*. The distinguishing characteristic of *solids* is that the attractive considerably exceeds the repulsive force. In solids, therefore, the cohesion is always considerable. The various properties of solids result from modifications of this molecular force.

11. *Tenacity.*—We find on trial that it is much easier to pull asunder a rod of lead than a rod of steel of the same thickness; showing that the molecules of some solids cohere more strongly than those of others.

When a solid is thus pulled asunder, it is said to be *ruptured.* The power which a solid has of resisting rupture is called *tenacity.*

The relative tenacity of different solids is determined by finding how much force is required to pull asunder a rod of the same thickness of each of the solids. If it takes twice as much force to pull asunder a rod of one of the solids as of another, the first is said to have twice the tenacity of the second.

The relative tenacity of solids may be determined by means of a machine called a *dynamometer*. This name is made up of two Greek words, and means *force-measurer*.

Fig. 4.

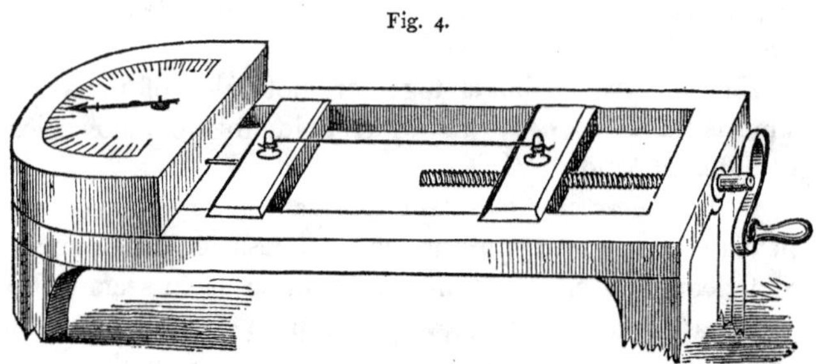

One form of the machine is represented in Figure 4. It consists of a heavy iron frame, at one end of which is a box containing a stout steel spring. A pointer connected with this spring moves over a graduated arc on the top of the box. On the frame are two movable blocks, or slides, one of which is attached to the spring, while the other may be carried backward and forward by means of a screw and crank.

The rod whose tenacity is to be tried is stretched between the two slides, and the crank is then cautiously turned so as to pull upon the rod until it breaks. The force which is thus brought to bear upon the rod bends the spring; and the position of the pointer when the rod breaks shows how much force was required to pull the rod asunder.

12. *Hardness and Softness.*—If we indent a piece of india-rubber with the finger-nail, or strike a piece of lead a smart blow with a hammer, we see that it is possible to dis-

place the molecules of a solid. When it is easy to displace the molecules, as in the case of wax, the solid is called *soft;* when it is difficult to displace them, as in the case of glass, the solid is called *hard.*

To find which of two solids is the harder, see which will scratch the other. The one which scratches is always harder than the one scratched. Diamond is the hardest solid known. Because of its hardness it is used for cutting glass, which is also a very hard substance.

13. *Elasticity, Brittleness, Ductility, and Malleability.* — When molecules have been displaced, one of three results must follow, — they will return to their original positions as soon as they are left to themselves, or they will take up permanently new positions, or they will fall entirely asunder.

If we bend a steel rod moderately, it straightens as soon as it is released; showing that the molecules sometimes tend to return to their former positions after they have been displaced. This tendency of the molecules to return to their original positions is called *elasticity.*

We find on trial that a rod of glass, or even of pipe-clay, will straighten on being released after being slightly bent. Every solid has been found to be elastic.

A steel rod may be bent a good deal, and yet straighten when released; but if it be bent beyond a certain point it will no longer straighten, showing that the molecules, after they have been displaced beyond a certain limit, no longer tend to go back to their original positions. The greatest extent to which the molecules of a solid can be displaced, and yet go back to their original positions, is called the *limit of elasticity* for that solid. While all solids are elastic, they differ very much in the limit of their elasticity. The molecules of steel and india-rubber can be displaced a good deal and yet return to their original positions, while those of glass and pipe-clay can be displaced but slightly.

If a glass rod be bent within a certain limit it will straighten when released, but if it be bent beyond this limit it will not remain permanently bent, but will break; showing that the molecules of a solid cannot always take up permanently new positions. When the molecules cannot take up permanently new positions, the solid is said to be *brittle.* Hard solids are likely to be brittle also; but hardness and brittleness are, as we have seen, entirely different things.

When the molecules of a solid can take up permanently new positions, it is ordinarily described as *malleable* or *ductile.* It is said to be *malleable* when it can be hammered or rolled out into sheets; *ductile* when it can be drawn out into wires.

Gold is one of the most malleable of the metals. In the manufacture of gold leaf it is hammered out into sheets so thin that it takes from 300,000 to 350,000 of them to make the thickness of a single inch.

The gold is first rolled out into sheets by passing it many times between steel rollers in what is called a rolling-machine. The rollers are so arranged that they can be brought nearer and nearer to each other, pressing the gold into a thinner and thinner sheet every time it is passed between them. After it has thus been rolled out to the thickness of writing-paper, it is cut up into pieces about an inch square. These are piled into a stack with alternate pieces of tough paper, and beaten with wooden mallets. They are again cut up into small pieces and arranged in a stack with alternate squares of gold-beater's skin, and again beaten with mallets. This last process is usually repeated three times.

Wire is made by drawing a rod of metal through a series of conical holes in a hardened steel plate. Each hole is a little smaller than the preceding, so that the rod becomes lengthened and diminished in thickness as it is drawn through one after another. A machine for drawing iron wire is represented in Figure 5. It consists of a reel on which the coarser wire is wound, a *drawing-plate* through

Fig. 5.

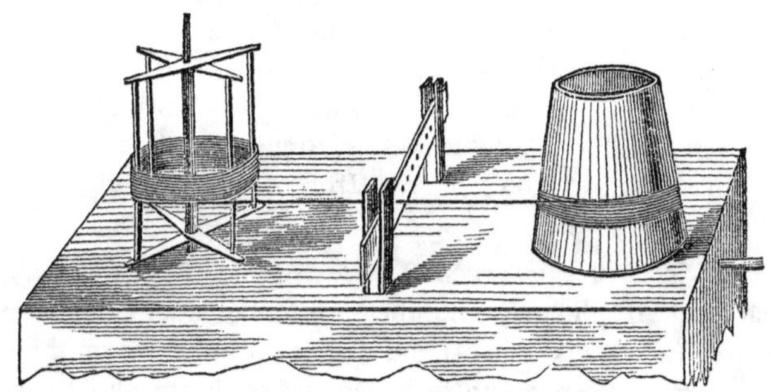

which it is pulled, and a drum on which it is wound again. The drum is turned by wheel-work, which is out of sight under the table.

The drawing of iron wire is attended with the following curious result. The molecules are separated in the drawing, yet the tenacity of the iron is greatly increased, so that fine iron wire is the most tenacious of substances. A bar one inch square of the best wrought-iron will sustain a weight of thirty tons; a bundle of wires one tenth of an inch in diameter, containing the same quantity of material, will sustain a weight of from thirty-six to forty tons; and if the wires have a diameter of only one twentieth or one thirtieth of an inch, the same quantity will sustain from sixty to ninety tons. Hence cables made of fine iron wire twisted together are much stronger than bars or chains of the same weight. The cables of suspension bridges are made in this way.

The following table gives the most useful metals in the order of their tenacity, malleability (both under the hammer and the rolling-mill), and ductility:—

Tenacity.	Malleability under the Hammer.	Malleability under the Rolling-Mill.	Ductility.
Iron	Lead	Gold	Platinum
Copper	Tin	Silver	Silver
Platinum	Gold	Copper	Iron
Silver	Zinc	Tin	Copper
Zinc	Silver	Lead	Gold
Gold	Copper	Zinc	Zinc
Lead	Platinum	Platinum	Tin
Tin	Iron	Iron	Lead

14. *Solids are somewhat Compressible.* — Pieces of oak, ash, or elm, plunged into the sea to the depth of 2,000 metres (about 6,560 feet) and drawn up after two or three hours, have been found to be compressed into about half their former bulk. Some metals are permanently diminished in bulk by hammering; and so also by the pressure to which they are subjected in the process of coining. The stone columns of buildings are frequently shortened by the great weight resting upon them. This was found to be the case with the columns supporting the dome of the Pantheon at Paris.

15. *The Arrangement of the Molecules.* — If alum be added to hot water as long as it will dissolve, and then the water be allowed to cool slowly, a part of the alum will be deposited on the bottom of the dish, not in a confused mass, but in beautiful and symmetrical forms. Such symmetrical forms are called *crystals*. If saltpetre, nitrate of baryta, or corrosive sublimate be treated in the same way, beautiful crystals will be formed, but in each case the crystals will have a different shape.

If sulphur be melted in a crucible, and then allowed to cool slowly till a crust forms on the surface, on carefully breaking the crust and pouring off the remaining liquid the crucible will be found lined with delicate needle-shaped crystals. In the same way crystals of bismuth and many of the metals may be obtained.

Fig. 6.

The cohesive force, then, not only holds the molecules of a solid together, but when it is free to act it often arranges these molecules in regular order, building them up into forms of great beauty.

In the cases of the formation of crystals which we have already described, the solid is first brought to the liquid state, and then allowed slowly to return to the solid state again. The solid was first reduced to a liquid that the molecules might have freedom of motion. The building of a crystal out of molecules is much like building a house out of bricks. The bricks must be taken one by one and laid in regular order before they are cemented together. So in forming a crystal, the molecules must be arranged one by one in regular order before they are fastened together by the cohesive force.

Large crystals of many solids can be obtained by dissolving as much of the solid as is possible in cold water, and then setting it away in a shallow dish where it will be free from dust and disturbance, and allowing the water to evaporate very slowly. The more gradual the formation, the larger are the crystals. The large crystals seen in cabinets of minerals were probably centuries in forming. The water in which the solid was dissolved found its way into a cavity of a rock and there slowly evaporated.

The tendency of the cohesive force to form the molecules into crystals is strikingly shown in cannon which have been many times fired, and in shafts of machinery and axles of car-wheels which are continually jarred. Such bodies often become brittle, and on breaking show the smooth faces of the crystals which have been formed. The continued jarring gives the molecules a slight freedom of motion, and crystals are slowly built up.

Many solids are crystalline in structure which do not appear to be so. Thus a piece of ice, as we shall prove elsewhere, is a mass of the most perfect crystals, but they are so closely packed together that we cannot readily distinguish them. There is a large class of solids, however, as the fats, which cannot be crystallized.

16. *The Molecules cohere more strongly on some sides than on others.* — It is easy to cleave a piece of mica in one direction, but difficult to cleave it in other directions. The same is true of all crystals. It is much easier to cleave them in certain directions than in others. This is also the case with some substances which are not crystalline, as wood, which splits readily in one direction only. These facts prove that the molecules cohere more strongly on some sides than on others. Iron and other solids are not so tenacious when crystalline in structure as when not crystalline. This is because the molecules in crystals are arranged in layers, so that the weakest sides are brought face to face.

17. *Annealing and Tempering.* — If melted glass be dropped into cold water, it forms the well-known *Rupert's drops*, which are so brittle that, if we break off the small end or scratch them slightly with a file, they fly in pieces. When glass is allowed to cool in the air at the ordinary temperature, it is also very brittle. In order to make it tough enough for ordinary use, it must be cooled very slowly. This slow cooling of glass or other substances is called *annealing*. Glass is annealed by passing it slowly through a long oven, which is kept very hot at one end and cool at the other. It is usually about two days in passing through the oven.

Steel, also, when suddenly cooled from a high temperature, is very hard and brittle, but when slowly cooled it is very tough and pliable. The process of bringing steel to the various degrees of hardness requisite for its uses in the arts is called *tempering*. Steel is usually tempered in the following manner. It is first heated white hot, and then suddenly cooled by plunging it into cold water. It is thus rendered very brittle. It is then reheated and allowed to cool slowly. When it is to be made quite hard, it is reheated but slightly; when quite soft, it is reheated a good deal. The more it is reheated, the softer it becomes on cooling. These different conditions of glass and steel are probably owing to differences in the arrangement of the molecules.

18. *Liquids.* The distinguishing characteristic of liquids is that the attractive and repulsive forces acting between the molecules are very nearly balanced, the attractive force being slightly in excess. Hence in liquids the cohesion is slight, and the molecules are free to move among themselves.

If a piece of lead be carefully measured, then melted and measured again, it will be found to have increased in bulk. The same is true of nearly all solids. Hence, when any substance is in a liquid state, the molecules are farther apart than when it is in a solid state. This explains why, in moulding bullets, the mould is never quite filled by the bullet.

There are, however, a few marked exceptions to this rule. It is well known, for instance, that if a bottle be filled with water and tightly corked, and allowed to freeze, the bottle will be burst. This shows that the molecules of ice are farther apart than those of water.

19. *Liquids are but slightly Compressible.* — The apparatus represented in Figure 7 consists of a very thick vessel of glass closed at top and bottom. Within the vessel are a piston which can be moved by the thumb-screw at the top, and a glass bulb which is prolonged by a very fine tube bent as represented. Fill the bulb and tube with any liquid,

as water, and plunge the end of the tube in the mercury which covers the bottom of the vessel. Then fill the vessel with water, and apply pressure by turning the screw. The mercury will rise in the tube, showing that the liquid in the bulb has been compressed. This compression, however, is but slight, amounting at most to a few millionths of the bulk of the liquid.

Fig. 7.

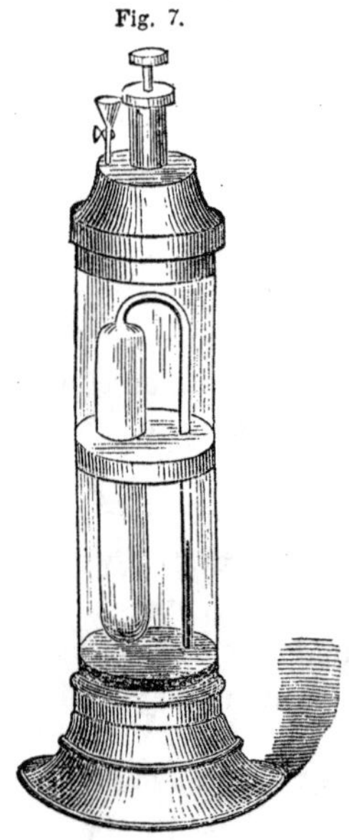

20. *Liquids are perfectly Elastic.* — However much the screw, in the above experiment, may be turned down, or however long it may be left, on loosening it the mercury will at once fall inside the tube to a level with the mercury outside; showing that liquids are perfectly elastic. This elasticity is, however, developed only when the liquid is compressed, that is, when the molecules have been brought nearer together. In whatever other way the molecules may be displaced, they show no tendency to return to their former positions.

21. *The Arrangement of Molecules in Liquids.* — If a mixture of water and alcohol be made so as to be just as heavy as sweet-oil bulk for bulk, and a quantity of the oil be carefully introduced into the centre of this mixture by means of a dropping-tube, the oil will neither rise nor sink, but gather into a beautiful sphere. This experiment shows that when the molecules of a liquid are left to themselves, they at once collect into spheres. What hinders the molecules of liquids from always taking this spherical form will be explained further on.

Rain-drops, dew-drops, and the manufacture of shot illustrate this tendency of the molecules of liquids. In the manufacture of shot, melted lead is poured through a sieve at the top of a very high tower, and the drops in falling take the form of spheres, which become solid before they reach the bottom.

22. *Gases.* — In gases, as has already been shown, the repulsive molecular force exceeds the attractive. Hence there is no cohesion in this state of matter, and the molecules move among themselves with greater freedom than those of liquids.

The molecules of any substance are farther apart in the gaseous state than in either the solid or liquid state. This may be shown by filling a test-tube nearly full of water, then closing it tightly with a cork through which a fine tube passes nearly to the bottom of the test-tube. On boiling the water so as to convert a portion of it into steam, which is a gas, the water is driven forcibly out of the fine tube; showing that the steam occupies more space than the water from which it comes.

Fig. 8.

Fig. 9.

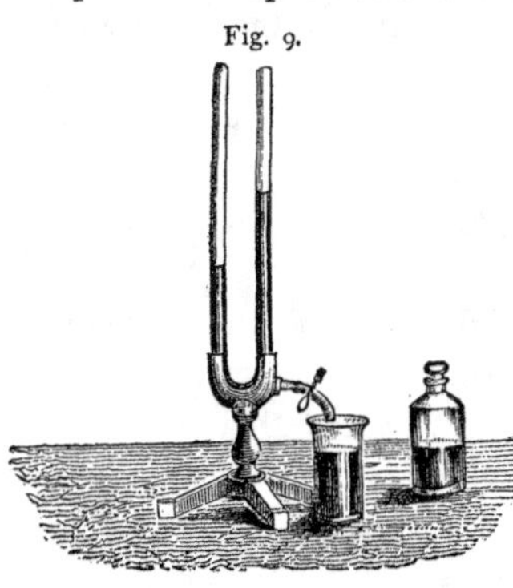

23. *Gases are readily Compressible.* — The figure represents a U-tube closed at one end and open at the other, with a nipper-tap* at the bend. Pour in mercury enough to cover the bend. The closed end is now filled with air. Pour in more

* See Appendix, I.

mercury, and this column of air rapidly shortens. The same would be true if the closed end of the tube were filled with any other gas; showing that gases are highly compressible.

24. *Gases are perfectly Elastic.*—Open the nipper-tap that the mercury may run out, and it is entirely driven out of the closed arm of the tube. To prove that it is the elasticity of the air which drives out the mercury from this arm, fill the closed arm and a part of the open arm with mercury, and open the nipper-tap. The mercury will flow out from the open arm, and not from the closed arm.

SUMMARY OF COHESION.

Matter is made up of definite but insensible masses, called *molecules.* (1, 2.)

These molecules are not in actual contact with one another. (3.)

It is probable that the spaces which separate the molecules are immense in comparison with the size of the molecules themselves. (4.)

By trying to pull a solid in two we learn that there is an *attractive molecular force*, which holds the molecules together.

By placing a rubber bag partially filled with a gas under the receiver of an air-pump, and exhausting the air, we find that there is also a *repulsive molecular force*, which pushes the molecules apart. (5.)

Since the molecules of a solid may separate on being heated, and yet hold firmly together, we conclude that these two molecular forces *act together*, and that the *repulsive* molecular force is *increased by heat.* (6.)

We find that there are *three states of matter*, depending upon the relative strength of these two forces: the *solid* state, in which the attractive force is considerably the greater; the *liquid* state, in which the two forces are nearly equal; and the *gaseous* state, in which the repulsive force is the greater. (7.)

The force which holds together molecules of the *same* kind is called *Cohesion;* that which holds together molecules of *different* kinds, *Adhesion.* (8.)

Cohesion is the excess of the attractive over the repulsive molecular force. In *solids*, it is comparatively strong; in *liquids*, it is weak; in *gases*, it does not exist.

The properties of solids depend on the action of the cohesive force. (10.)

The *tenacity* of a solid is its power of resisting rupture. (11.)

A solid is called *hard* when it is difficult to displace its molecules; *soft*, when it is easy to displace them. (12.)

Elasticity is the tendency of the molecules, on being displaced, to return to their original positions. All solids are elastic, but differ greatly in the *limit* of their elasticity.

A solid is said to be *brittle* when its molecules cannot take up permanently new positions.

It is said to be *malleable* or *ductile* when they can take permanently new positions: *malleable*, when it can be hammered or rolled into sheets; *ductile*, when it can be drawn into wire. (13.)

Solids are somewhat *compressible.* (14.)

The cohesive force often arranges the molecules of a solid into regular forms, called *crystals.* (15.)

Crystals can be split more easily in some directions than in others, showing that the *cohesive force is stronger on some sides of the molecule* than on others. (16.)

The molecules are farther apart in the *liquid* than in the solid state; yet liquids are *less compressible* than solids. (19.)

Liquids are *perfectly elastic;* but their elasticity is developed only when the molecules are brought nearer together. (20.)

The molecules of a *liquid*, when acted upon only by cohesion, tend to collect into *spheres.* (21.)

In the *gaseous* state, the molecules are farther apart than in the liquid state. (22.)

Gases are readily *compressible*, and when compressed are *perfectly elastic.* (23, 24.)

II.

ADHESION.

ADHESION.

25. *Adhesion between Solids and Solids.* — Adhesion has already been defined as the force which holds together unlike molecules.

The sticking of the chalk to the blackboard, of the graphite of the pencil to paper, and of dust to furniture, prove the existence of this force between solids and solids. The use of the various cements also illustrates this force, and also the fact that some solids adhere to a given solid more strongly than others. If we wish to fasten two pieces of wood together, we use glue; if two bricks or stones, we use mortar, or some calcareous cement; if two pieces of glass, sealing-wax, or some resinous substance. Stone adheres to mortar more strongly than wood or glass, and wood adheres to glue more strongly than stone or glass.

When solids are held together by cements, cohesion and adhesion are both brought into play. When, for instance, two pieces of wood are held together by means of glue, the adhesive force holds the wood on each side to the glue, and cohesion holds together the molecules of the glue.

When furniture breaks, we often see that the wood splits instead of separating from the glue. So also stones are sometimes cemented together so firmly that the stone itself will break sooner than separate from the cement. These facts show that the adhesive force between two solids is frequently stronger than the cohesive force of the solids themselves.

26. *Adhesion between Solids and Liquids.*—If we dip the hand in water it comes out wet. This, and similar facts equally familiar, prove that there is also an adhesive force between liquids and solids.

27. *The Adhesion between a Liquid and a Solid is sometimes not strong enough to overcome the Cohesion of the Liquid.*—If a glass disc be suspended from one pan of a balance and counterpoised by weights, and then brought in contact with

Fig. 10.

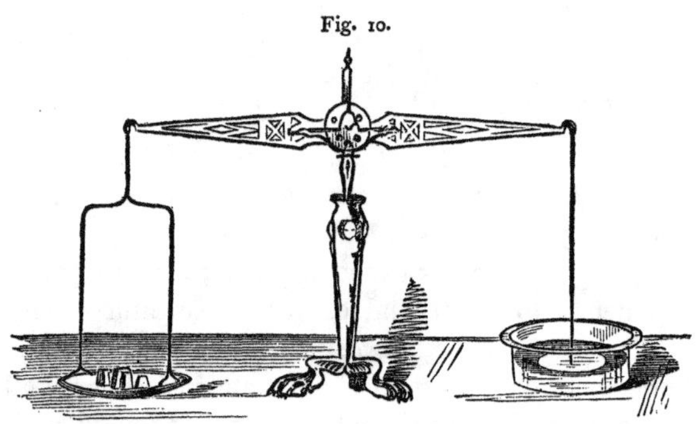

mercury, it will require additional weight to raise the disc from the mercury, and the disc comes off dry. This proves, first, that there is adhesion between glass and mercury, and, secondly, that this adhesion is not strong enough to overcome the cohesion of the mercury.

28. *The Adhesion between a Solid and a Liquid is sometimes strong enough to overcome the Cohesion of the Liquid.*—If a glass plate be laid upon the surface of water and then removed, it comes off wet, that is to say, covered with a film of water; showing that the adhesion between a solid and liquid is sometimes strong enough to overcome the cohesion of the liquid.

Since adhesion takes place only at the surface, it is evident that we may increase the adhesion of a solid for a liquid by increasing the surface of the solid.

If we take any solid, as a stone, and break it in two, the stone evidently has all the surface it had before it was broken, and, in addition, the two surfaces exposed by the breaking. Hence the more a solid is broken up, the more surface it exposes. The readiest way, then, to increase the surface of a solid is to pulverize it.

If pulverized bone-black be mixed with ordinary vinegar, or with wine, and the liquid be separated again by pouring the mixture upon a piece of unsized paper placed inside a funnel, every trace of color will be removed from the liquid. All vegetable colors can be removed from liquids in the same way. Removing the color from a liquid in this way is called *clarifying* the liquid.

Bone-black is obtained by burning bones in closed vessels. It is pulverized that it may present more surface. Other substances are sometimes used for clarifying liquids. Next to bone-black, or "animal charcoal," as it is sometimes called, ordinary charcoal is the best and the most frequently used. The bone-black evidently removes the coloring matter by means of the adhesive force which exists between the two. It is also evident that the coloring matter adheres to the bone-black more strongly than to the liquid, else the two would not be separated.

Use is made of this property of charcoal in the refining of sugar. The dark-colored syrup which is obtained from the cane is first filtered through long sacks filled with coarsely pulverized charcoal. In this way all the color is removed. The colorless syrup is then evaporated and forms white sugar.

29. *The Adhesion between a Solid and a Liquid is sometimes strong enough to overcome the Cohesion of the Solid.*—If some Epsom salts be put into water, the salts will speedily be reduced to the liquid state. The adhesive force between the water and the salts has evidently overcome the cohesive force of the solid, since it has reduced the solid to the liquid state.

30. *Summary.* — We have, then, three well-marked cases of adhesion between solids and liquids : —

1st. When the adhesive force is not strong enough to overcome the cohesion of the liquid. In this case the liquid *cannot wet* the solid.

2d. When the adhesive force is strong enough to overcome the cohesion of the *liquid.* In this case the liquid can *wet* the solid.

3d. When the adhesive force is strong enough to overcome the cohesion of the *solid.* In this case the liquid can *dissolve* the solid. The liquid which dissolves the solid is called a *solvent,* and the liquid in which the solid has been dissolved is called a *solution.*

31. *Heat promotes Solution.*—We find on trial that Epsom salts will dissolve more rapidly and in greater quantity in hot than in cold water. This is as we should expect, since we have already found that heat tends to overcome cohesive force. As a general rule, solids dissolve in greater quantities and more readily in hot than in cold liquids, but there are exceptions.

32. *Different Solids are not equally soluble in the same Liquid.*—If we put a piece of sealing-wax into water, it does not dissolve at all ; while water will dissolve about twice its bulk of Epsom salts. If we compare other solids in the same way, we shall scarcely find any two which dissolve with equal readiness in the same liquid.

33. *The same Solid is not equally soluble in different Liquids.* — Sealing-wax, which does not dissolve in water, dissolves quite readily in alcohol. This shows that the same solid does not dissolve with the same readiness in different liquids.

34. *Capillarity.* — If one end of a fine and clean glass tube be put into water, the water will be seen to rise inside the tube above the surface of the water outside. If one end of a similar tube be put into mercury, the mercury will

be seen to fall inside the tube below the surface of the mercury outside. This action of liquids inside tubes is called *capillarity*. The force which draws some liquids into tubes and pushes others out, has been called *capillary force*. This name is a convenient one, and we shall retain it, though, as we shall show elsewhere, capillarity results from the combined action of certain other forces.

We have seen already that water will *wet* glass, while mercury will *not*. We have, then, two well-marked cases of capillarity, corresponding to two cases of adhesion between solids and liquids; for those liquids which will wet a tube are drawn into it, while those which will not wet it are driven out. Mercury will wet zinc, and it is drawn into a tube of zinc, just as water is into a tube of glass.

We find by using glass tubes of different sizes, that the finer the tube, the higher the water rises and the lower the mercury falls, that is, the more marked is the capillarity. This fact explains the name. The word *capillary* comes from a Latin word (*capillaris*) which means *hair-like*. The force was called *capillary* because its action is most powerful in hair-like tubes. This force, however, acts in tubes of every size, and in fact a tube is not necessary for its action; as may be seen by putting two plates of glass together as represented in the figure, and then dipping them into water or mercury. The water will rise between the plates, and the mercury fall.

Fig. 11.

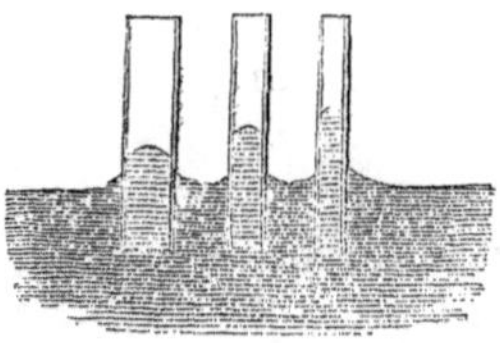

Fig. 12.

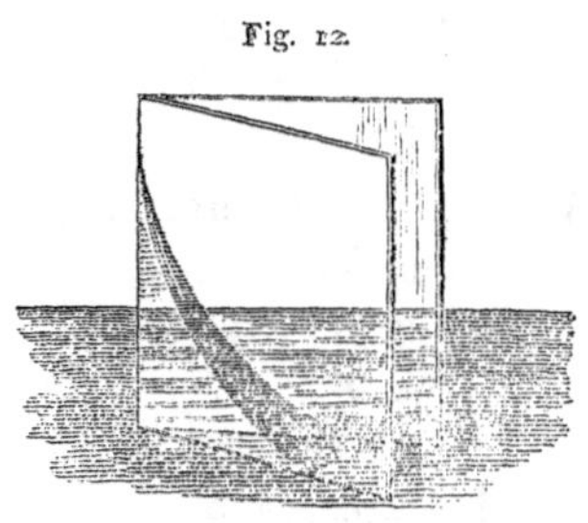

35. *Illustrations of Capillarity.* — A lamp-wick is full of

tubes and pores, and capillary force draws the oil up through these to the top of the wick, where it is burnt. When one end of a cloth is put into water, capillary force draws the water into the tubes and pores of the cloth, and the whole soon becomes wet. In the same way a lump of sugar, or other porous substance, soon becomes wet throughout, if a corner of it is put into water. Blotting-paper is full of pores into which the capillary force draws the ink. The use of a towel for wiping anything which is wet depends on the same principle.

36. *Strength of the Capillary Force.* — It is well known that, when a piece of cloth is wet, it is almost, if not quite, impossible to wring or squeeze it dry. This shows that the capillary force which holds the water into the pores of the cloth is very strong. Some solids, as wood, swell on becoming wet. If holes are drilled into a granite rock, and dry wooden plugs driven into them, and water poured over the ends of the plugs, the capillary force draws the water into the pores of the wood, which swells and splits the rock. This is a striking illustration of the strength of the capillary force.

37. *Capillary Force never causes a Liquid to flow through a Tube.* — If a glass tube be so fine that the capillary force will draw water into it to the height of two inches, and the tube be then lowered so that not more than half an inch shall be above the surface of the water, the water will not overflow the tube. If, however, the water be removed as soon as it comes to the top, more will rise in the tube to take its place.

When a lamp is burning, the oil is passing up continually through the wick, because it is burned as soon as it reaches the top; but when the lamp is not burning, the oil does not overflow the wick. The wick of an alcohol lamp must be covered with a cap when the lamp is not burning, otherwise the alcohol will evaporate as fast as it comes to the top of the wick, and so all pass out of the lamp.

38. *Adhesion between Solids and Gases.* — If a small glass jar inverted over mercury be filled with ammonia gas,* and a piece of boxwood charcoal, previously heated to redness and cooled by plunging it into the mercury, be introduced into the jar, the mercury rapidly rises into the jar, and, if the piece of coal is large enough, entirely fills it. The ammonia gas, then, has been drawn into the charcoal by an adhesive force; proving the existence of adhesion between the molecules of a solid and those of a gas. When a gas is taken up in this way by any substance, it is said to be *absorbed.*

If we try other solids instead of charcoal, we shall find that no two absorb ammonia gas with equal readiness.

If a piece of boxwood charcoal be introduced into a jar of air inverted over mercury, the mercury rises in the jar very slowly; showing that different gases are not absorbed with equal readiness by the same solid.

When the ammonia gas is absorbed by the charcoal, as in the above experiment, it evidently occupies less space than before. When a gas is absorbed by a solid, then, the repulsive force of the gas has to be overcome. We have already seen that cold helps to overcome the repulsive force. Hence we should expect that a solid would absorb a gas when cold more readily than when hot. Experiment shows this to be true.

Heat, on the other hand, increases the repulsive force. If a solid which has absorbed a gas be heated, the repulsive force of the gas is increased, so that it finally overcomes the adhesion of the solid for the gas, which then leaves the solid. The charcoal is heated before it is introduced into the jar, in order to drive all the air out of its pores.

39. *Adhesion between Liquids and Liquids.* — If oil be poured upon water, the oil, which is the lighter, soon rises

* See Appendix, 2.

to the top and remains entirely separate from the water. If, however, alcohol, which is also lighter than water, be poured into water, the two will thoroughly mix. This may be made evident to the eye by using colored water.

The fact that the alcohol remains mixed with the water proves that the molecules of the alcohol must adhere to those of the water, and that this adhesion is strong enough to overcome the cohesion of the liquids.

Nearly all liquids will mix when poured together, though some will mix much more readily than others.

40. *Diffusion of Liquids.* — If some colored alcohol be put into a tall glass jar, and then, by means of a funnel and a long tube reaching to the bottom of the jar, some water be carefully poured in, the water will remain a short time at the bottom of the jar, and its separation from the alcohol will be sharply defined. On standing a few days, the liquid will become of the same color throughout; showing that the alcohol and water have mixed. This mixing of liquids on being merely brought into contact is called *diffusion of liquids.* Different liquids diffuse into each other at very different rates, while some, as oil and water, will not diffuse at all.

Fig. 13.

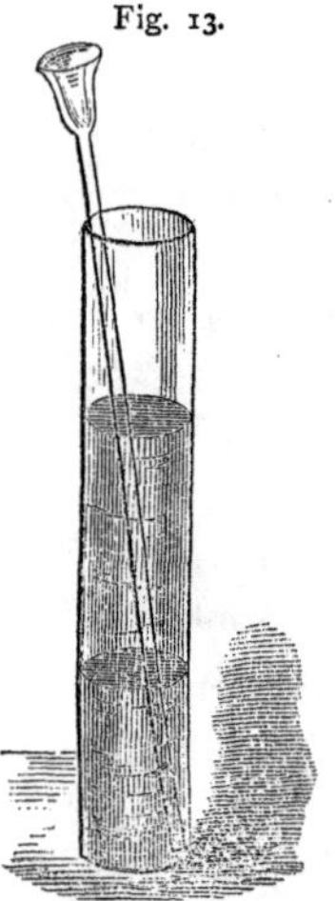

41. *Osmose of Liquids.* — If a bladder be fastened air-tight to the end of a long glass tube, and the bladder be filled with alcohol and introduced into a vessel of water, the liquid will gradually rise in the tube; showing that the water has passed into the bladder. At the same time the alcohol passes slowly out and mixes with the water.

The mixing of liquids when separated by a thin membrane or porous substance is called *osmose of liquids.*

Liquids do not mix at the same rate when separated by a thin membrane or porous substance as when they mix by simple diffusion. The rate of mixing is modified in a striking manner by the presence of the membrane or porous substance. If, in the above experiment, we substitute a small collodion balloon for the bladder, the liquid will fall in the tube, showing that now the alcohol passes out more rapidly than water passes in.

Fig. 14.

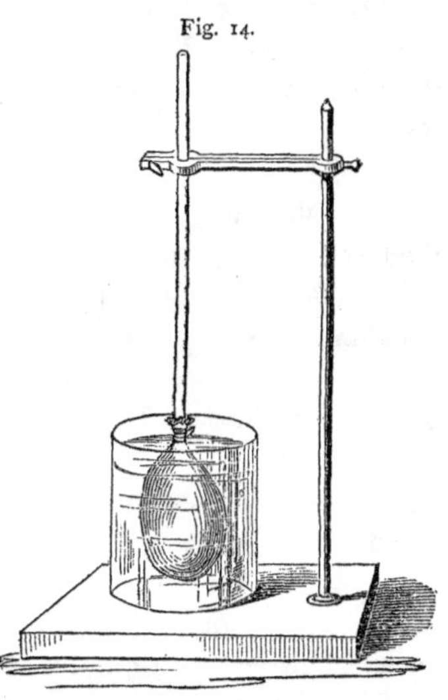

The cases of osmose already given have been explained as the combined effect of capillarity and diffusion. In the first case, the water, which will wet the bladder much more readily than alcohol, is drawn by capillary force into the minute pores of the bladder, and thus is carried to its inner surface, where, coming in contact with the alcohol, it diffuses into it. As the water is thus removed as fast as it comes to the inner surface, a constant flow is maintained through the pores of the bladder. At the same time the alcohol passes out slowly by diffusing through the water which fills the pores of the bladder. In the second case, the alcohol wets the collodion more readily than the water does; hence the more rapid flow is in the opposite direction.

Capillarity and diffusion, which explain these cases of osmose so satisfactorily, will by no means explain all cases. No satisfactory explanation has yet been given of the various ways in which the presence of the membrane affects the rate of mixing.

42. *Adhesion between Liquids and Gases.* — Let a small glass jar inverted over mercury be filled with ammonia gas, and then some water be poured over the surface of the mercury. If now the jar be carefully raised, the moment the mouth of the jar comes in contact with the water, the latter rises and completely fills the jar; showing that the ammonia has been absorbed by the water, and consequently that the molecules of the gas adhere to those of the water.

The same gas is not absorbed with equal readiness by all liquids, as is shown by the fact that ammonia gas, which is absorbed so greedily by water, is not absorbed at all by mercury. The same liquid does not absorb all gases with equal readiness, as is shown by inverting a jar of air over water. The air is absorbed scarcely at all.

43. *Cold and Pressure promote Absorption.* — When a gas is absorbed by a liquid, as well as when absorbed by a solid, the molecules are brought nearer together and the repulsive force overcome. Both cold and pressure, as we have seen, help to overcome this force; hence they favor the absorption of a gas by a liquid.

The effect of pressure on the absorption of a gas by a liquid is illustrated in soda-water. Soda-water owes its agreeable taste mainly to the presence of carbonic acid gas in the water. Water and carbonic acid are brought into contact in the fountain, and subjected to very great pressure. When the water is drawn from the fountain this pressure is removed, and the carbonic acid which had been taken up by the water escapes in thousands of little bubbles, causing the liquid to foam, or *effervesce.*

Ordinary liquid ammonia, or *aqua ammonia*, is a gas absorbed by water. If this liquid be heated, the ammonia gas escapes. The heat increases the repulsive force of the gas, and thus enables it to overcome its adhesive force for the liquid. The ordinary way to free a liquid from an absorbed gas is to heat it.

Common spring-water owes much of its pleasant taste to the presence of carbonic acid and other gases which it absorbs from the air. When this water is boiled, these gases escape, and it becomes very insipid. The constant agitation of running water helps it to absorb gases, since it is thus made to present more surface to the air.

44. *Diffusion of Gases.* — Two bottles are connected by a long glass tube. The lower bottle is then filled with carbonic acid and the upper with hydrogen gas, which is very much lighter than carbonic acid.* After a time the hydrogen will be found to have passed down and mixed with the heavier carbonic acid, and the carbonic acid to have mixed with the hydrogen in the upper bottle.

Fig. 15.

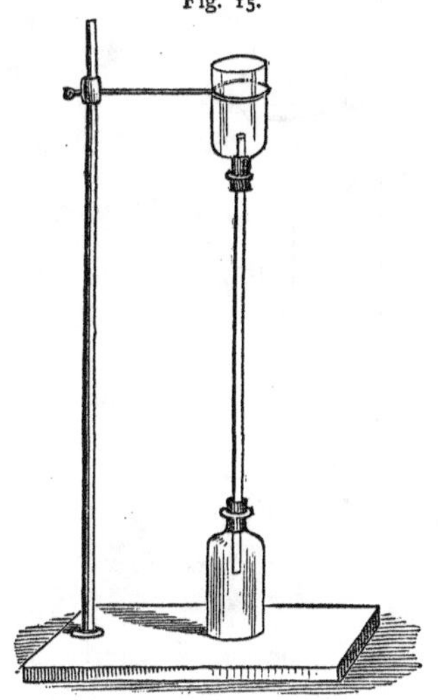

The presence of the carbonic acid in the upper bottle may be proved by pouring into it lime-water,† which on shaking becomes of a milk-white color. Hydrogen has no effect upon lime-water.

The mixing of gases when brought into contact is called *diffusion of gases.*

Different gases diffuse into each other at very different rates. As a general thing, the more the gases differ in weight, the more rapidly they diffuse into each other.

45. *Osmose of Gases.* — A long glass tube is fastened air-tight, by means of a cork and sealing-wax, into the open

* See Appendix, 3.

† See Appendix, 4.

end of an unglazed porcelain cup such as is used in a Bunsen's or Grove's battery. The cup is then held so that the end of the tube dips beneath the surface of water, and a large bell-jar of hydrogen* is held over the cup. There is an instant rush of bubbles from the end of the tube up through the water, showing that the hydrogen has passed through the pores of the cup and mixed with the air inside. Remove now the jar of hydrogen, and the water at once rises in the tube, showing that the hydrogen inside the cup has passed out through the pores to mix with the air outside.

Fig. 16.

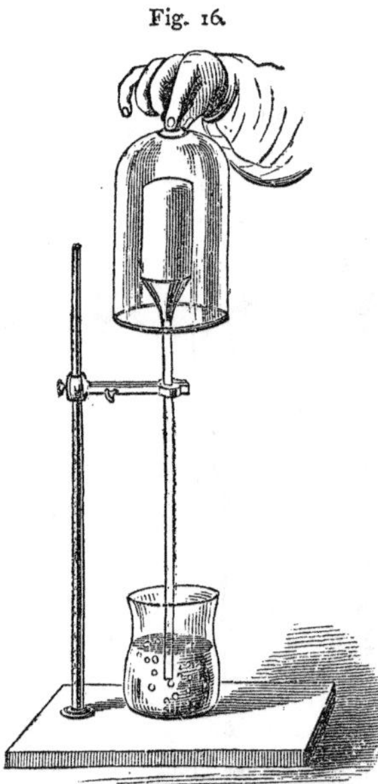

The mixing of gases when separated by a porous substance or thin membrane is called *osmose of gases.*

The diffusion and osmose of gases point to the existence of adhesion between molecules of different gases; but the existence of this adhesion has not been fully established.†

* See Appendix, 5.

† See Appendix, 6.

SUMMARY OF ADHESION.

Adhesion is the force which holds together molecules of *different kinds.*

It acts between molecules of solids and solids, solids and liquids, solids and gases; also between liquids and liquids, and liquids and gases. It is doubtful whether there is any adhesion between the molecules of different gases.

The adhesive force between two solids is sometimes greater than the cohesive force of the solids themselves. (25.)

There are three cases of adhesion between solids and liquids: —

1st. When the adhesion is not strong enough to overcome the cohesion of the liquid, and the liquid *cannot wet* the solid.

2d. When it is strong enough to overcome the cohesion of the liquid, and the liquid can *wet* the solid.

3d. When it is strong enough to overcome the cohesion of the solid, and the liquid can *dissolve* the solid. (26 – 30.)

Heat generally promotes solution, since it helps to overcome the cohesion of the solid. (31.)

The same liquid dissolves some solids more readily than others; while some liquids dissolve the same solid more readily than others do. (32, 33.)

Capillary force is a force acting upon liquids within tubes.

Liquids which can wet a tube are drawn into it by the action of this force, while liquids which cannot wet it are driven out of it.

The finer the tube, the more marked is the capillarity. (34.)

The capillary force is a very strong force; but acting

alone it never causes a liquid to flow *through* a tube. (36, 37.)

When a gas is absorbed by a solid or by a liquid, the adhesive force between the molecules of the solid or liquid and those of the gas must be strong enough to overcome the repulsive force of the gas. (38, 42.)

Heat hinders absorption, since it increases the repulsive force between the molecules of the gas. Hence gases absorbed by solids or liquids can be separated from them by means of heat. (38, 43.)

The same solid or liquid absorbs some gases more readily than others; while the same gas is absorbed by some solids or liquids more readily than by others. (38, 42.)

The adhesive force between the molecules of different liquids causes the liquids to mix. The mixing of liquids on merely coming in contact with each other is called *diffusion* of liquids. (40.)

Liquids also mix when separated by a thin membrane or porous substance. This mixing is called *osmose* of liquids. (41.)

Gases, like liquids, mix by *diffusion* and by *osmose.* (44, 45.)

III.

CHEMICAL AFFINITY.

CHEMICAL AFFINITY.

46. *Action of Potassium and Sodium on Water.*—If a piece of potassium be thrown upon water, it burns with a rose-colored flame, and rapidly disappears. A piece of sodium when thrown upon water moves briskly about with a hissing sound, and soon disappears, but without flame. When, however, the sodium is kept still by putting it upon blotting-paper which lies upon the surface of the water, it burns with a yellow flame.

Fig. 17.

In both cases the metal disappears, and the water apparently remains unchanged. But if a strip of red litmus-paper be put into the water on which the sodium or potassium was burned, it turns at once to a blue color, showing that the water has also been changed; for ordinary water has no effect on red litmus-paper.

If a piece of sodium be put into a small metallic cup pierced with holes and provided with a handle, and the cup be quickly introduced bottom upward under the mouth of a test-tube previously filled with water and inverted over a vessel of water with its mouth dipping beneath the surface, bubbles of gas rise from the cup and soon fill the test-tube.

If now the test-tube be raised from the water and a lighted taper applied to its mouth, the gas takes fire with a slight explosion, and burns with a pale bluish flame.

Fig. 18.

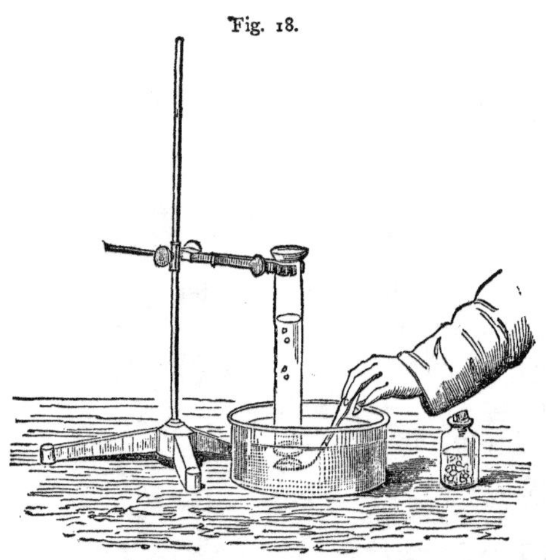

This gas is called *hydrogen.* Its extreme lightness, and the peculiar way in which it takes fire and burns, serve to distinguish it from other gases. Its lightness may be shown by the following simple experiment. Fill a test-tube with hydrogen and hold it mouth downward for some time; then apply a lighted taper, and it will still be found full of hydrogen. Fill it again and hold it mouth upward, and after a little time apply a lighted taper, when the hydrogen will be found to have wholly escaped. It is so light that it rises through the air as a cork rises through water.

47. *Whence does the Hydrogen come?* — In Figure 19 we have a U-tube open at both ends. In each of its arms is put a small graduated glass bell-jar. At the bend of the tube directly under each jar a platinum wire passes through the glass. This wire is flattened on the inside and terminates in a hook on the outside. The whole apparatus is

filled with water, and one of the platinum wires is connected with the zinc pole, and the other with the platinum pole of a Grove's battery.* As soon as these connections are made, bubbles rise from each platinum wire, and the bell-jars slowly fill with gas.

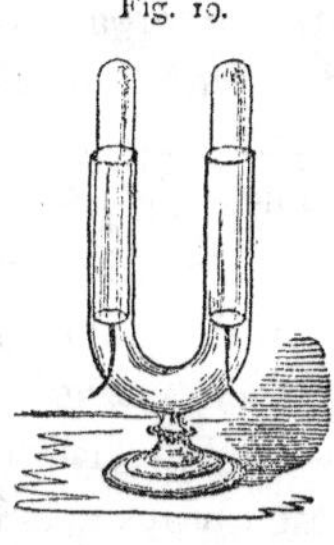
Fig. 19.

When the jars are full, remove the one which has been connected with the zinc pole, and bring a lighted taper to its mouth. The gas takes fire with a slight explosion, and is recognized at once as hydrogen.

Remove the other jar, hold it mouth upward, and plunge into it a splinter of wood with a spark of fire on its end. The splinter at once bursts into a flame, and the gas does not burn. This gas is called *oxygen.*

We conclude, then, that the hydrogen in the previous experiment came from the water, and that it was set free by the sodium; and that besides hydrogen there is another gas in water called *oxygen.*

48. *Are these the only Substances found in Water?* — If a jet of hydrogen be burned in a jar filled with oxygen,† and the heated gases be conducted through a cold glass tube, moisture collects on the sides of the tube, trickles down, and drops from the end, and may be caught in a wine-glass. This liquid resembles water; and if we drop a little of it on a piece of potassium, the latter takes fire and burns with a rose-colored flame, showing that the liquid really is water. This water can have been formed from nothing else than hydrogen and oxygen, hence there can be nothing else in water.

49. *Compound Substances and Elements.* — We see, then, that water can be separated, or *decomposed,* into two gases, hydrogen and oxygen. Substances which, like water, can

* See Appendix, 7.

† See Appendix, 8.

be decomposed into other substances, are called *compound* substances.

It has hitherto been found impossible to decompose either hydrogen or oxygen. Substances which, like oxygen and hydrogen, cannot by any known process be decomposed into simpler substances, are called *elements*.

Only sixty-five elements are known, and thirteen out of these sixty-five, combining in various ways, make up almost the whole of the substances found in and about the earth.

50. *Chemical Force.*—The force which causes the elements to combine is called *chemical force* or *affinity*. Chemical force always acts between unlike substances, and in this respect it resembles adhesion; but it differs from it in two or three important particulars:—

1st. When water adheres to a glass plate, neither the properties of the water nor those of the glass are changed. So, too, when sugar is dissolved in water, the sugar still retains its sweet taste and all its peculiar characteristics. It is simply sugar reduced to the liquid state and diluted with water. But when chemical force causes hydrogen and oxygen to combine, they form *water*, a substance wholly unlike hydrogen, which is inflammable, and oxygen, which aids combustion.

The first characteristic of chemical force is that it changes the properties of the substances which it causes to combine. It not only unites different substances, but it unites them so as to form a new substance. It *combines* them.

2d. If water be again decomposed in the U-tube by means of the battery, it will be noticed that the tube into which the hydrogen rises fills just twice as fast as the other. In whatever way water is decomposed, we always get just twice as much hydrogen as oxygen *by measure*.

Figure 20 represents a U-tube, one arm of which is closed, the other open, with a nipper-tap at the bend. Two platinum wires pass through the closed arm near the top and almost meet within, while their outer ends are formed into loops for the attachment of battery wires.

Introduce into the closed end of this tube a mixture of hydrogen and oxygen, using first just twice as much hydrogen by measure as oxygen. Combine the gases by means of the electric spark.* Water is of course produced, and all the gases are used up, as is shown by the mercury's rising to the top of the tube.

Fig. 20.

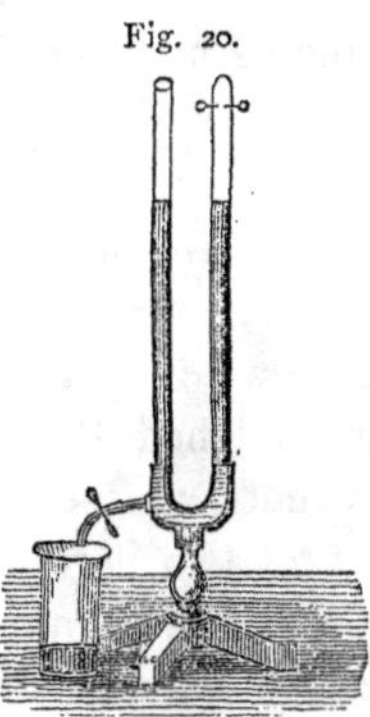

If another mixture is used in which we have more than two measures of hydrogen to one of oxygen, some hydrogen will be left over. If less than two measures of hydrogen to one of oxygen be used, some oxygen will be left. So that in whatever proportions hydrogen and oxygen be mixed, we find that when they form water two measures of hydrogen always combine with one of oxygen.

We may dissolve a teaspoonful of sugar in a wine-glass or in a pitcher of water, and there will still be sugar in every part of the water. Adhesion acts between quantities of matter altogether indefinite.

The second characteristic of chemical force is that it always acts between *definite quantities* of matter.

Chemical force, then, may be briefly described as the force which *acts between definite quantities of different kinds of matter, and causes them to combine to form a new substance.*

51. *Chemical Force is stronger between some Substances than between others.*—When potassium is thrown upon water (46), it unites with the oxygen and sets the hydrogen free; showing that the chemical force between potassium and oxygen is stronger than between hydrogen and oxygen. In

* See Appendix, 9.

general, if to a compound of two elements, A and B, a third element, C, be added, which has a stronger affinity for A than A has for B, A will leave B and combine with C.

Any change which is occasioned by chemical force is called a *chemical change*.

ATOMS AND ATOMIC WEIGHTS.

52. We have already found that two measures of hydrogen always combine with one measure of oxygen to form water. If equal measures of hydrogen and oxygen are weighed, the oxygen is found to weigh sixteen times as much as the hydrogen. There are, then, in water, by *weight*, two parts of hydrogen to sixteen parts of oxygen.

When ice is melted, then, it is resolved into molecules, each of which contains 2 parts by weight of hydrogen and 16 parts of oxygen. When acted upon, at ordinary temperatures, by either sodium or potassium, 1 part of this hydrogen is removed by 23 parts of the former, or 39 parts of the latter, giving rise to compounds each molecule of which contains 1 part of hydrogen, 23 parts of sodium, and 16 parts of oxygen, or 1 of hydrogen, 39 of potassium, and 16 of oxygen. The first of these compounds is called *hydrate of sodium*; the second, *hydrate of potassium.* On heating the first with sodium, the remainder of the hydrogen is displaced by 23 additional parts of sodium, and a compound formed each molecule of which contains 46 parts of sodium and 16 of oxygen. On heating the second with potassium, the remainder of the hydrogen is displaced by 39 parts of potassium, and a compound formed containing 78 parts of potassium and 16 of oxygen.

When chlorine (an element with which we shall soon become acquainted) is passed through a red-hot tube along with steam, it removes the hydrogen from the water and sets the oxygen free.* It is found that chlorine, like sodium

* See Appendix, 10.

and potassium, can displace either one half or the whole of the hydrogen, giving rise to two compounds. It is, however, impossible to replace any fraction of the oxygen of water with chlorine. No substance has yet been formed which will displace any other fraction than one half of the hydrogen from water; and no substance is known which will displace any fraction of the oxygen without displacing the whole.

We see, then, that by means of affinity a molecule of water can be divided into three parts, 2 of which are hydrogen and 1 is oxygen; and that the 1 part of oxygen weighs 16 times as much as each of the 2 parts of hydrogen. No way is known of further dividing this molecule.

It has also been found that no substance can displace less hydrogen from any of its compounds than sodium or potassium, and that it always requires 23 parts of the one and 39 parts of the other to displace one part of hydrogen.

We therefore conclude that the smallest parts into which hydrogen, sodium, and potassium can be divided by affinity have the relative weights of 1, 23, and 39; that is, the smallest part into which sodium can be divided by affinity weighs 23 times as much as the smallest part into which hydrogen can be divided by the same means; and the smallest part into which potassium can be divided weighs 39 times as much as the smallest part into which hydrogen can be divided.

These smallest parts into which substances can be divided by affinity are called *atoms*, from a Greek word meaning *indivisible*, and the relative weights of the atoms of the elements are called *atomic weights*. Thus 1, 23, 39, and 16 are the atomic weights of hydrogen, sodium, potassium, and oxygen respectively.

53. *Masses, Molecules, and Atoms.* — We have, then, a threefold divisibility of matter: —

1st. Matter can be divided by mechanical means into minute but *sensible* portions, or *masses*.

2d. Matter can be divided by means of heat into *insensible* portions, called *molecules.*

3d. Matter can be divided by means of chemical force into portions which are by us *indivisible,* called *atoms.*

SYMBOLS.

54. To indicate briefly the composition of a substance, and also such changes as we have already seen brought about by the chemical force, a system of notation by *symbols* has been devised.

55. *Symbols of Elements.*—The symbol for an element is always the first letter of its name (in some cases its Latin name), unless the names of two or more elements begin with the same letter, in which case a second letter is added to distinguish them.

Thus the symbol for hydrogen is H; for oxygen, Θ; for potassium (Latin *kalium*), K; for sodium (*natrium*), Na; for phosphorus, P; etc.

The symbol of an element, when used alone, indicates one atom of that element.* Thus H stands not only for hydrogen, but also for one atom, or one part by weight, of hydrogen;

Θ	for	1 atom or	16	parts	by weight	of	oxygen;
K	"	"	39	"	"	"	potassium;
Na	"	"	23	"	"	"	sodium;
P	"	"	31	"	"	"	phosphorus.

Many of the symbols of the elements have a bar across them, for a reason which will be explained hereafter.

56. *Symbols of Compounds.*—The symbol for a compound is obtained by joining together the symbols of the elements contained in it, writing after the symbol of each

* The symbols are, however, sometimes used merely as abbreviations of the names of the elements; H for hydrogen in general, K for potassium, etc.

element a figure to indicate how many atoms of that element are contained in a molecule of the compound.

Thus the symbol for water is $H_2\Theta$; the symbol for oxide of potassium is $K_2\Theta$; the symbol for oxide of sodium is $Na_2\Theta$; for hydrate of potassium, $HK\Theta$; for hydrate of sodium, $HNa\Theta$.

57. *The Meaning and Use of the Symbols.* — The symbol $H_2\Theta$ indicates that water is a compound of hydrogen and oxygen; that in each molecule of water there are two atoms of hydrogen and one atom of oxygen, and that by weight there are 2 parts of hydrogen to 16 parts of oxygen. The symbol $HNa\Theta$ indicates that hydrate of sodium is composed of hydrogen, sodium, and oxygen; that in a molecule of this compound there is one atom of each of these elements; and that in this substance there is one part by weight of hydrogen to 23 parts of sodium and 16 parts of oxygen.

By means of these symbols chemical changes are very concisely indicated. Thus the change which takes place when potassium is thrown upon water can be indicated thus:

$$H_2\Theta + K = HK\Theta + H.$$

The sign $+$ (plus) indicates merely that the substances are together without combination.

The change which takes place when all the hydrogen in water is displaced by potassium is indicated by symbols thus:

$$H_2\Theta + 2K = K_2\Theta + 2H.$$

When the symbol of an element stands by itself, and we wish to express two or more atoms of it, the figure is placed *before* the symbol instead of after it. Thus we write 2K rather than K_2.

When hydrogen and oxygen combine to form water (50), the change is indicated by symbols thus:

$$2H + \Theta = H_2\Theta.$$

By means of these symbols we can readily determine what fraction of the weight of the whole compound the weight of each element in it forms. For instance, we wish to find what fraction of the weight of water is hydrogen, and what fraction is oxygen. The symbol for water is $H_2\Theta$. H_2 indicates 2 parts by weight, and Θ indicates 16 parts. In water, then, there are 18 parts by weight, 2 of which are hydrogen and 16 oxygen: $\frac{2}{18}$ or $\frac{1}{9}$ of the weight of water is hydrogen, and $\frac{16}{18}$ or $\frac{8}{9}$ is oxygen.

What fraction of the weight of hydrate of sodium is hydrogen, what fraction is sodium, and what fraction oxygen?

The symbol for this compound is $HNa\Theta$.

H = 1 part by weight of hydrogen;
Na = 23 parts " " sodium;
Θ = 16 " " " oxygen;
40 = whole number of parts by weight in $HNa\Theta$. Then,

$\frac{1}{40}$ of $HNa\Theta$ is hydrogen;
$\frac{23}{40}$ " " " sodium;
$\frac{16}{40}$ " " " oxygen.

It will now be easy to solve such problems as the following: —

How much hydrogen is there in 45 kilogrammes of water?

We have found that $\frac{1}{9}$ of the weight of water is hydrogen. In 45 kilogrammes of water, then, there are 5 kilogrammes of hydrogen.

How many kilogrammes of water could be produced by using 64 kilogrammes of oxygen?

We have found that $\frac{8}{9}$ of water is oxygen. The question then becomes, if $\frac{8}{9}$ of the weight is 64 kilogrammes, what is $\frac{9}{9}$, or the whole? Evidently 72 kilogrammes.

How many kilogrammes of oxygen would be required to convert 92 kilogrammes of sodium into oxide of sodium? How many kilogrammes of oxygen and of hydrogen, to convert it into hydrate of sodium?

$\frac{23}{31}$ of $Na_2\Theta$ is sodium; $\frac{8}{31}$ is oxygen. If $\frac{23}{31} = 92$ kilogrammes, $\frac{1}{31} = 4$ kilogrammes, and $\frac{8}{31} = 32$ kilogrammes $=$ weight of oxygen required to convert the sodium into oxide of sodium.

Again, $\frac{1}{40}$ of $HNa\Theta = H$; $\frac{23}{40} = Na$; $\frac{16}{40} = \Theta$. If $\frac{23}{40} = 92$ kilogrammes, then $\frac{1}{40} = 4$ kilogrammes $=$ weight of H required, and $\frac{16}{40} = 64$ kilogrammes $=$ weight of Θ required.

58. *Problems.* — 1. Indicate by symbols the chemical change which takes place when sodium is thrown upon water.

2. Indicate by symbols the change which takes place when the hydrogen of water is wholly displaced by sodium.

3. What fraction of the weight of oxide of potassium is oxygen, and what fraction is potassium?

4. What fraction of the weight of hydrate of potassium is hydrogen, what potassium, and what oxygen?

5. How much water can be formed by using 75 grammes of oxygen? How much hydrogen would be required?

6. How much potassium and how much oxygen in 235 grammes of oxide of potassium?

7. How much hydrogen would be required to make 3 kilogrammes of hydrate of potassium? To make 3 kilogrammes of hydrate of sodium?

8. If 892 grammes of oxygen were used in making oxide of sodium, how much sodium would be used?

9. If the same amount of oxygen were used in making hydrate of potassium, how much potassium would be used?

MURIATIC ACID.

59. We will next examine a liquid which somewhat resembles water in appearance, and which, though not so well known, has for a long time been extensively used in the arts. This liquid is commonly called *muriatic acid,* a

name which, like *water*, gives no indication of its composition. Its scientific name will be given when we have found out of what the acid is composed.

Put some muriatic acid into a flask, and connect the flask by means of a tube with a glass jar filled with mercury and inverted over the mercury trough, as shown in

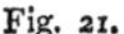

Fig. 21.

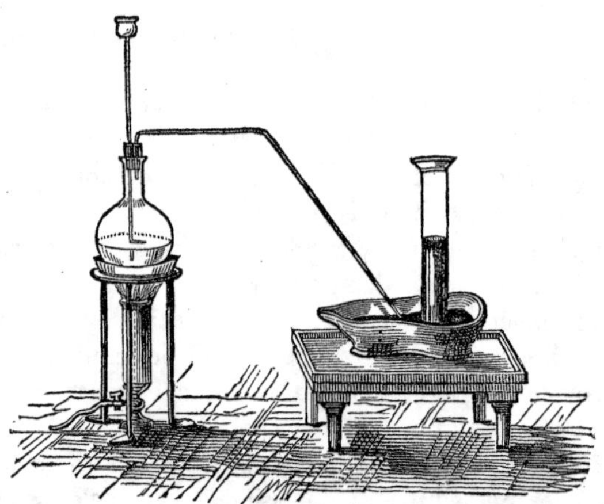

the figure.* Boil the acid, and a gas will pass over and fill the jar. If now the jar be removed and its mouth be opened under water, the gas will be quickly absorbed, and the water will have all the properties of muriatic acid.

We therefore conclude that ordinary muriatic acid is a gas reduced to the liquid state by being absorbed by water. It is thus made liquid merely for convenience in using and transporting it.

When this muriatic acid is boiled, steam will be generated and pass over with the gas.

We have seen that a solid absorbs some gases much

* See Appendix, 11.

more readily than it absorbs others. If we can find a solid which will absorb steam and not absorb muriatic acid gas, we evidently can separate these two gases by causing them to pass together over this solid. *Fused chloride of calcium* is such a solid. We break it up into coarse lumps, that it may expose the more surface to the gases, and put it into a glass tube or a tall and narrow glass jar. Such a tube or jar is called a *drying-tube* or *drying-jar.*

If some muriatic acid be again boiled in a flask, and the muriatic gases which escape be conducted, as shown in the

Fig. 22.

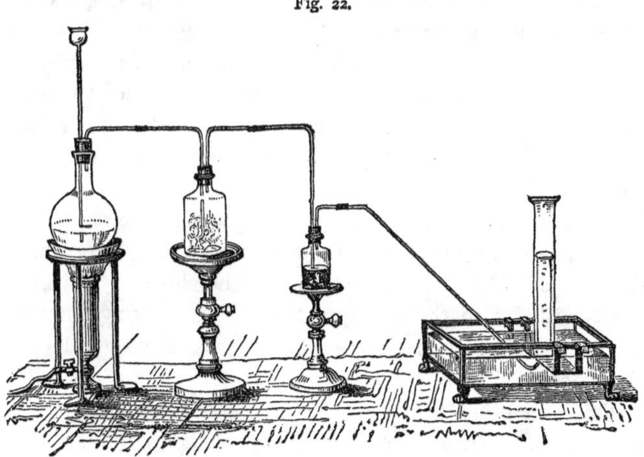

figure, first through a drying-jar, to remove the steam, and then through a bottle partially filled with *sodium amalgam* (sodium dissolved in mercury), and then into a jar filled with water, and inverted over the water-trough, the gas which collects in this jar will be found to be wholly unlike muriatic acid gas. It is no longer absorbed by water, and has no effect on blue litmus-paper. Muriatic acid gas is not inflammable, but if this gas be tested by bringing a lighted taper to the mouth of the jar after raising it from the trough, it takes fire with a slight explosion, and burns with a pale blue flame. This shows the gas to be hydrogen.

60. *Composition of Muriatic Acid.* — We see then that muriatic acid is a compound substance, and that one of the elements of which it is composed is hydrogen. We must next find what element or elements the sodium has removed from the acid. The mercury in the amalgam takes no part in the decomposition of the gas. The sodium is dissolved in the mercury merely that it may present more surface to the gas.

Taking the same apparatus used for decomposing water by means of the battery,* fill it with dilute muriatic acid, and connect it with the battery as before. The bell-jar connected with the zinc pole is filled quite rapidly with a gas which is found on trial to be hydrogen. The other jar after some time begins to fill with a gas of a yellowish-green color.

61. *Chlorine.* — This gas has a peculiarly suffocating odor, and the remarkable property of bleaching vegetable colors. This latter characteristic may be shown by raising the jar filled with the gas, inverting it, and putting into it a piece of moistened litmus-paper, which very soon loses all its color.

This gas from its color is called *chlorine;* a name derived from a Greek word meaning *yellowish-green.* Its color and its bleaching properties serve to distinguish it from other gases.

Chlorine may be more readily obtained from muriatic acid by mixing it with black oxide of manganese and heating the mixture gently in a flask.†

If a jar of chlorine be inverted over water, it is slowly absorbed. This is the reason that, when muriatic acid is

* See Appendix, 12. † See Appendix, 13.

decomposed by means of the battery, the chlorine does not at first collect in the bell-jar.

62. *Are Hydrogen and Chlorine the only Elements in Muriatic Acid?* — Take two glass jars of the same size, with ground mouths, fill one with hydrogen and the other with chlorine, and place them together mouth to mouth, as represented in Figure 23. Then withdraw the glass plates by which they are closed, shake the jars to mix the gases, and open them over a burning lamp. The mixed gases take fire, and the appearance of a white cloud shows that muriatic acid is formed by their combustion.* Hence this acid can contain only hydrogen and chlorine.

Fig. 23.

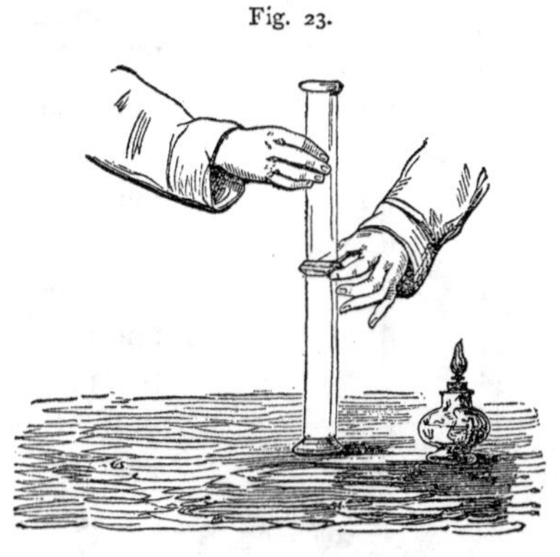

Fig. 24.

63. *How much Hydrogen and how much Chlorine in Muriatic Acid?* — Figure 25 represents a glass tube divided by a glass stop-cock into two unequal parts, and closed at its ends by glass stoppers. Fill the smaller division of the tube with dry chlorine and the larger one with dry hydrogen; then open the stop-cock and expose the tube for some

* If a dish of ammonia is standing near by, the white fumes will be much more dense.

hours to diffused daylight, after which bring it into direct sunlight. Now open one end of the tube under water, and the water will rise and fill a space double that which was occupied by the chlorine. The gas which remains is found to be hydrogen.

Fig. 25.

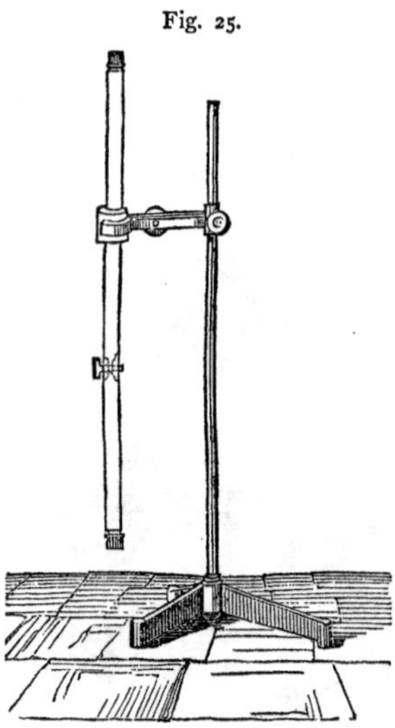

Fill now the smaller division of the tube with hydrogen, the larger one with chlorine, and let the two combine under the influence of light as before. The water will rise in the tube to the same height as before, but the gas which remains will be found to be chlorine.*

In these two experiments it is evident that equal measures of hydrogen and chlorine have combined; the excess, whether of the one or the other gas, remaining unchanged. By measure then, hydrogen and chlorine combine in equal parts to form muriatic acid.

We find on trial that chlorine weighs 35.5 times as much as hydrogen, bulk for bulk. By weight, then, there is one part of hydrogen to 35.5 parts of chlorine in muriatic acid.

64. *Atomic Constitution of Muriatic Acid.* — It is found impossible to displace any fraction of either the hydrogen or the chlorine from muriatic acid without displacing the whole. We must hence conclude that in a molecule of muriatic acid there is one atom of hydrogen and one of chlorine, and that the atomic weight of chlorine is 35.5.

* See Appendix, 14.

The symbol for chlorine is Cl. That of muriatic acid evidently will be HCl.

65. *Problems.* — 10. In 35 grammes of muriatic acid gas how much hydrogen and how much chlorine?

11. How much muriatic acid can be made by using 13 kilogrammes of hydrogen? By using 22 kilogrammes of chlorine?

12. How much hydrogen is required to convert 46 kilogrammes of chlorine into muriatic acid? How much chlorine to convert 46 kilogrammes of hydrogen into muriatic acid?

AMMONIA.

66. The ammonia of commerce, like the muriatic acid, is a gas absorbed by water, as may readily be shown by the method described under muriatic acid. (59.)

Fig. 26.

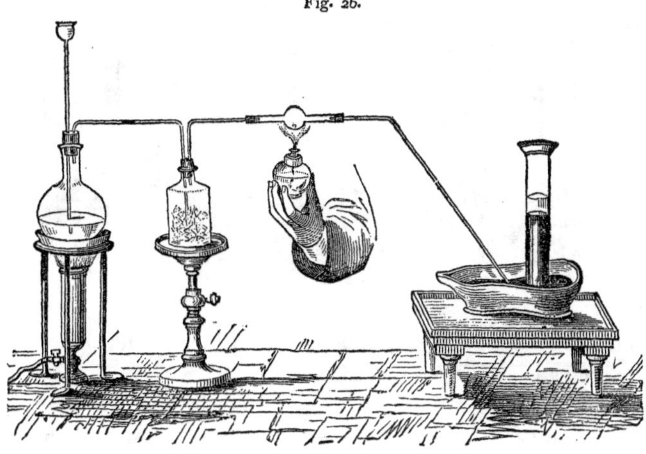

If some ammonia be boiled in a flask, steam and ammonia gas will pass off. If these gases are then passed through a drying-jar filled with unslaked lime, the steam will be retained, and the ammonia gas will pass on. If this gas be then conducted into a test-bulb in which there is a piece of heated potassium, and the gas which leaves the tube be collected in a small jar,* it will be found by the usual test to be hydrogen.

The apparatus required for this experiment is shown in Figure 26.

Ammonia is then a compound substance, one of whose elements is hydrogen.

67. *What Element has been removed by the Potassium?*— Fill a tolerably large bottle partly full of strong ammonia,

Fig. 27.

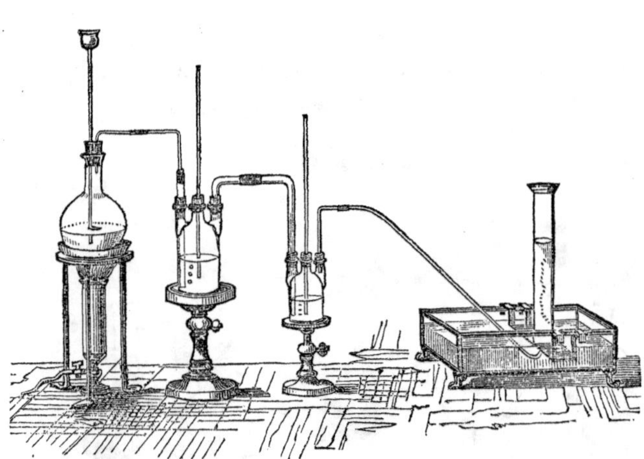

and close it with a cork through which pass two tubes, one reaching to the bottom of the bottle, and the other passing through the cork. Let a stream of chlorine pass through the first tube. As it enters the ammonia, flashes of light

* See Appendix, 15.

are seen, and other indications of energetic action. The gas which escapes from the bottle through the other tube is passed through a bottle filled with water (called a wash-bottle), and is then collected in jars. This gas is evidently not ammonia, for it was not absorbed by the water. It is not chlorine, for it is colorless. It is not hydrogen or oxygen, for it neither burns when a lighted taper is applied to it, nor will a lighted taper burn in it. It does not affect either blue or red litmus-paper. It cannot be decomposed, and must therefore be regarded as an element.

It is called *nitrogen*, and its symbol is N.

68. *Is there any other Element in Ammonia?* — Hydrogen and nitrogen cannot be made to combine directly. We cannot therefore employ this method to determine whether ammonia gas contains more than these two elements. But if a gramme of ammonia gas be decomposed into hydrogen and nitrogen, the weight of the hydrogen and nitrogen together will be just one gramme. Hence it is clear that there is nothing but these two elements in ammonia.

Fig. 28.

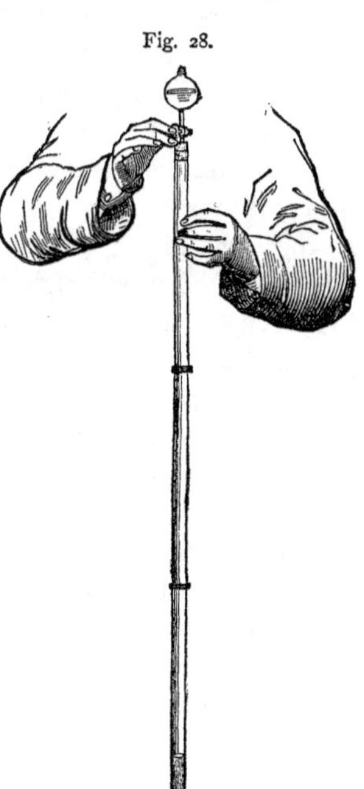

69. *How much of each of these Elements in Ammonia?* — Let a long tube (Fig. 28), closed at one end and divided into three equal parts by marks on the side, be filled with chlorine, and a

dropping-tube (see Fig. 29) filled with ammonia be connected with it by means of a rubber cork so as to form an air-tight joint. If now the ammonia be allowed to drop slowly into the tube, the chlorine will decompose it, uniting with the hydrogen and setting the nitrogen free. After enough ammonia has passed in to fill the tube to the depth of about half an inch, allow dilute sulphuric acid to pass into the tube by means of the dropping-tube as long as it will. It will fill the tube just two thirds full. The other third will be filled with a gas which is found on trial to be nitrogen.

Fig. 29.

While this nitrogen has been set free, how much hydrogen has been taken away? The chlorine combines with the hydrogen to form hydrochloric acid. In this acid there are equal measures of hydrogen and chlorine. As the tube was full of chlorine at first, enough hydrogen to fill the tube must have been taken from the ammonia. By measure, then, there are three parts of hydrogen and one of nitrogen in ammonia.

Nitrogen weighs 14 times as much as the same bulk of hydrogen. Hence by weight there are in ammonia 3 parts of hydrogen to 14 parts of nitrogen.

70. *Atomic Constitution of Ammonia.* — It is also found that, when ammonia is decomposed by chlorine, three definite compounds can be obtained; one containing one third less hydrogen than ammonia does, one containing two thirds less hydrogen, and one containing no hydrogen. Every time a measure of hydrogen is displaced, an equal measure of chlorine replaces it.

We therefore conclude that a molecule of ammonia contains three atoms of hydrogen and one atom of nitrogen; for none of the nitrogen can be displaced unless the whole is. The atoms of hydrogen can be successively displaced by atoms of chlorine. The symbol for ammonia, then, will be H_3N; and the atomic weight of nitrogen is 14.

71. *Problems.* — 13. What is the symbol for each of the three compounds that may be formed when ammonia is decomposed by chlorine?

14. What fraction of ammonia is hydrogen, and what fraction is nitrogen?

15. How much hydrogen and how much nitrogen in 186 grammes of ammonia?

16. How much ammonia can be made by using 15 grammes of hydrogen? By using 15 grammes of nitrogen?

MARSH GAS.

72. When the mud at the bottom of stagnant ponds and marshes is stirred, bubbles of gas are seen to rise. This gas is called *marsh gas.* It may readily be prepared by heating strong vinegar in a flask (of glass, or better of copper or iron), together with a mixture of lime and the caustic soda of commerce, and collecting the gas over water.*

Fig. 30.

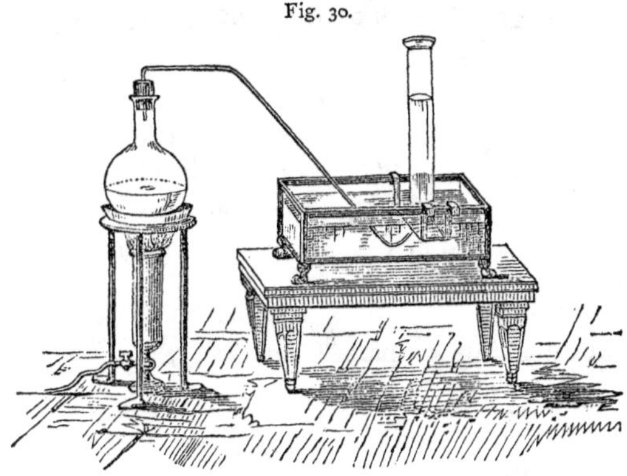

73. *Its Composition.* — If marsh gas be mixed with chlo-

* See Appendix, 19.

rine in a tall jar, and the mixture be ignited, a black powder will be deposited on the sides of the jar. This powder has been proved to be an element, and is called *carbon.* It is the same substance as charcoal. Its symbol is €.

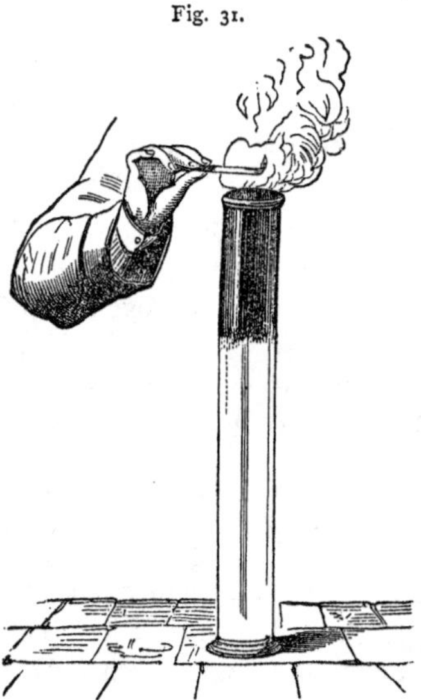

Fig. 31.

If a jet of marsh gas be burned in a jar of oxygen, moisture will collect on the sides of the jar, showing that water has been formed. Now water contains oxygen and *hydrogen.* The hydrogen must have come from the marsh gas which was burned. This gas, then, must contain at least two elements, hydrogen and carbon.

When a weighed quantity of marsh gas is decomposed, the hydrogen and carbon obtained from it weigh just as much as the gas decomposed; hence this gas can contain only hydrogen and carbon.

74. *Atomic Constitution of Marsh Gas.* — On careful analysis, marsh gas is found to contain 4 parts of hydrogen by weight to 12 parts of carbon. It is also found by experiment that fourths of the hydrogen can be successively displaced from marsh gas by chlorine, giving rise to four compounds; —

the 1st	containing	3	parts of H,	35.5	of Cl,	12 of €,
" 2d	"	2	" " H,	71	" Cl,	12 " €,
" 3d	"	1	" " H,	106.5	" Cl,	12 " €,
" 4th	"	0	" " H,	142	" Cl,	12 " €.

Hence we conclude that a molecule of marsh gas contains four atoms of hydrogen and one of carbon; and that the atomic weight of carbon is 12.

The atoms of hydrogen can be successively displaced by atoms of chlorine, giving rise to the four compounds mentioned above.

75. *Problems.* — 17. What is the symbol for marsh gas, and for each of the four compounds mentioned?

18. What fraction of the weight of marsh gas is hydrogen, and what fraction is carbon?

19. In 868 grammes of each of the four compounds mentioned, what weight of hydrogen, of chlorine, and of carbon?

COMBINING POWER OF THE ELEMENTS.

76. We see, then, that the atoms of different elements have different combining powers.

Thus one atom of chlorine has power to fix in combination only *one* atom of hydrogen, while one atom of oxygen has power to fix *two* atoms of hydrogen, one atom of nitrogen to fix *three* atoms of hydrogen, and one atom of carbon to fix *four* atoms of hydrogen.

We have also seen that the atoms of different elements can displace each other in compounds.

Thus one atom of potassium or sodium can displace *one* atom of hydrogen from water, forming hydrate of potassium or of sodium; that is,

$$H_2\Theta + K = HK\Theta + H.$$

But one atom of barium can displace *two* atoms of hydrogen from water. Thus,

$$H_2\Theta + Ba = Ba\Theta + 2H.$$

Whenever either sodium or barium can displace hydrogen from its compounds, one atom of sodium is always able to displace *one* atom of hydrogen, and one atom of barium *two* atoms of hydrogen.

It is also found that one atom of aluminium is able to displace *three* atoms of hydrogen from certain of its compounds, and one atom of tin to displace *four* atoms of hydrogen.

One atom of chlorine or of potassium, then, seems to have the same combining or "atom-fixing" power as *one* atom of hydrogen; one atom of oxygen or of barium, the same combining power as *two* atoms of hydrogen; one atom of nitrogen or of aluminium, the same as *three* atoms of hydrogen; and one atom of carbon or of tin, the same as *four* atoms of hydrogen.

77. *Monads, Dyads, Triads, and Tetrads.* — An element whose atom has the same combining or "atom-fixing" power as one atom of hydrogen is called a *monatomic* element, or, more briefly, a *monad;* names derived from the Greek word for *one.*

An element whose atom has the same combining or "atom-fixing" power as two atoms of hydrogen is called a *diatomic* element, or a *dyad;* from the Greek word for *two.*

An element whose atom has the same combining or "atom-fixing" power as *three* atoms of hydrogen is called a *triatomic* element, or a *triad;* from the Greek word for *three.*

An element whose atom has the same combining power as *four* atoms of hydrogen is called a *tetratomic* element, or a *tetrad;* from the Greek word for *four.*

Every element now known is supposed to belong to one of these four groups.

The combining or "atom-fixing" power of the hydrogen atom is regarded as unity. Then each atom of a *monad* will have *one* unit of combining power; one atom of a *dyad* will have *two* units of combining power; one atom of a *triad* will have *three* units of combining power; and one atom of a *tetrad* will have *four* units of combining power.

78. *Saturated Compounds.* — A compound is regarded as *saturated* when the combining power of all its atoms is satisfied. Thus carbonic acid (CO_2) and marsh gas (H_4C) are saturated compounds; since in the first the 4 combining units of carbon are satisfied by the 4 units of the two atoms of oxygen, and in the second the four units of carbon are satisfied by the 4 units of the 4 atoms of hydrogen. But carbonic oxyde (CO) and olefiant gas (H_2C) are non-saturated compounds; since in the first the 2 combining units of the oxygen satisfy only 2 of the 4 combining units of the carbon, and in the second only 2 of the 4 combining units of the carbon are satisfied by the 2 units of hydrogen.

The atom-fixing power of the different elements is sometimes indicated thus:—

C''''	indicates	that	1	atom of	carbon	has a combining power of	4 units,
N'''	"	"	1	"	nitrogen	"	3 units,
O''	"	"	1	"	oxygen	"	2 units,
H'	"	"	1	"	hydrogen	"	1 unit.

79. *Classification of the Elements.* — The following is a table of the most important elements arranged into groups according to Miller. The names of the elements are given along with their symbols and atomic weights. Those in the first division of each group are the so-called *non-metallic* elements; those in the second division of each group are *metals*.

MONADS.

Hydrogen	H	1
Fluorine	F	19
Chlorine	Cl	35.5
Bromine	Br	80
Iodine	I	127

Potassium (*Kalium*)	K	39
Sodium (*Natrium*)	Na	23
Silver (*Argentum*)	Ag	108
DYADS.		
Oxygen	Θ	16
Sulphur	S	32
Barium	Ba	137
Cadmium	Cd	112
Calcium	Ca	40
Chromium	Cr	52.5
Cobalt	Co	59
Copper (*Cuprum*)	Cu	63.5
Iron (*Ferrum*)	Fe	56
Lead (*Plumbum*)	Pb	207
Magnesium	Mg	24.3
Manganese	Mn	55
Mercury (*Hydrargyrum*)	Hg	200
Nickel	Ni	59
Strontium	Sr	87.5
Uranium	U	120
Zinc	Zn	65.5
TRIADS.		
Nitrogen	N	14
Phosphorus	P	31
Arsenic	As	75
Boron	B	10.9
Aluminium (*or* Aluminum)	Al	27.5
Antimony (*Stibium*)	Sb	122
Bismuth	Bi	210
Gold (*Aurum*)	Au	196.6
TETRADS.		
Carbon	C	12
Silicon	Si	28

Tin (*Stannum*)	Sn	118
Platinum	Pt	197

It will be noticed that there is a bar across the symbols of all the elements of the *dyad* and *tetrad* groups in the above table, while of those belonging to the *monad* and *triad* groups the symbol of Aluminium is the only one thus marked.

COMPOUNDS OF OXYGEN AND NITROGEN.

80. *Nitric Oxide.* — If one part by weight of saltpetre, 8 parts of copperas, and 8 parts of dilute sulphuric acid be gently heated in a large flask, a gas will be rapidly disengaged which, when collected in a jar over water, is colorless.

If in a porcelain cup floating on water a small piece of phosphorus be ignited, and a jar filled with oxygen be inverted over it, the phosphorus will burn with great brilliancy, and the jar will be filled with a dense white cloud.* This is soon absorbed by the water, which rises in the jar; showing that some of the oxygen has united with the phosphorus to form the white fumes.

If now we repeat the experiment, filling the jar with the gas just obtained, the phosphorus burns just as it did in the oxygen, and the same white fumes are formed. There must, then, be oxygen in this gas.

But the gas is not pure oxygen, since a lighted taper put into it is speedily extinguished.

Let a mixture of equal volumes of this gas and hydrogen be passed through a glass tube containing some platinized asbestos,† and the tube be heated to dull redness by means of a lamp. Hold a piece of red litmus-paper in the gas which escapes from the tube, and its color is at once

* See Appendix, 16. † See Appendix, 17.

changed to blue. Now ammonia is the only gas that changes red litmus-paper to blue. Hence, when the mixed gases pass over the heated asbestos ammonia is formed. The asbestos remains unchanged, and may be used for years. There must then be nitrogen in the gas which was mixed with the hydrogen and passed through the tube.

That there is nothing else than nitrogen and oxygen in this gas can be proved by the method already described in the case of ammonia and marsh gas (68, 73).

This composition of the gas is indicated by its name, *nitric oxide.*

81. *Nitrous Acid and Nitric Acid.* — If a jar be filled with nitric oxide and another jar of about half the capacity filled with oxygen be inverted over it,* the two gases on mixing will at once become of a bright cherry-red color. They evidently combine and form a new substance.

Boil a little aquafortis in a flask,† and let the steam pass through a porcelain tube heated to dull redness. The gas which comes from the tube will be of a bright red color, and will be at once recognized as the compound of nitrogen and oxygen with which we have just become acquainted.

Fig. 32.

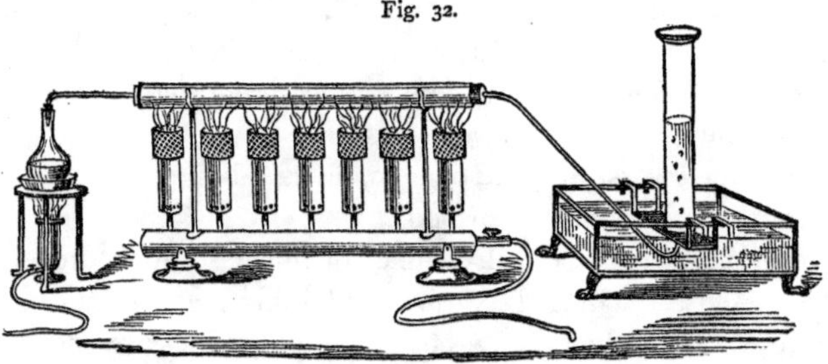

If the gases which issue from the porcelain tube be col-

* See Appendix, 18. † See Appendix, 19.

lected in a jar over water, the red gas will be absorbed by the water, and the jar will be filled with a colorless gas, which by the usual tests is found to be oxygen.

The steam of aqua-fortis is then decomposed in the heated tube into the red gas and oxygen. Hence aqua-fortis must be a compound of nitrogen and oxygen, since the red gas has been found to be a compound of those elements. The scientific name of aqua-fortis is *nitric acid*, and that of the red compound is *nitrous acid*.

82. *Atomic Constitution of the Compounds of Nitrogen and Oxygen.* — The three compounds, *nitric oxide*, *nitrous acid*, and *nitric acid*, which are so unlike in properties, differ in composition only in the proportion of oxygen which each contains.

Nitric oxide is converted into nitrous acid by taking up more oxygen, while nitric acid is converted into nitrous acid by losing a part of its oxygen. Hence nitrous acid contains more oxygen than nitric oxide does, and nitric acid more than nitrous acid.

In the compounds of hydrogen which we have examined we have found hydrogen combining with each element in only one proportion; here we find oxygen combining with nitrogen in several proportions, and forming compounds no less unlike one another than the compounds of hydrogen with different elements. It is one of the peculiarities of hydrogen that it seldom combines with an element in more than one proportion, or, in other words, that it seldom forms other than saturated compounds, while it is no less a peculiarity of oxygen that it usually combines with the same element in several proportions, or, in other words, that it readily forms non-saturated compounds (78). It combines with nitrogen in five different proportions, forming compounds whose symbols and names are as follows: —

$N_2\Theta$, Nitrous Oxide,
$N\Theta$, Nitric Oxide,

$N_2\Theta_3$, Nitrous Acid,
$N\Theta_2$, Hyponitric Acid,
$N_2\Theta_5$, Nitric Acid.

83. *Problems.* — 20. What three things does each one of these symbols indicate?

21. What fraction of the weight of each of these compounds is oxygen? What fraction of each is nitrogen?

22. How much oxygen and how much nitrogen in 756 grammes of each of these compounds?

23. How much of each of these compounds could be made by using 62 kilogrammes of nitrogen, and how much oxygen would be required in each case?

24. How much of each compound could be made by using 62 kilogrammes of oxygen, and how much nitrogen would be required in each case?

ACIDS, BASES, AND NEUTRALS.

84. When phosphorus is burned in oxygen, the jar is filled with a white cloud, which is soon absorbed by the water (80).

If now a piece of red litmus-paper be dipped in the solution, its color will not be changed. If, however, a piece of blue litmus-paper, whose color is not affected by a solution of oxide of sodium, be dipped in this solution, its color is at once changed to red.

It will be remembered that the solution of the hydrate of sodium formed by replacing the hydrogen of water with sodium (46) has the power to turn red litmus-paper to blue. It will also be remembered that water has no effect on either red or blue litmus-paper.

We see, then, that oxygen combines with certain elements, as potassium and sodium, to form a compound whose solution changes red litmus-paper to blue; with certain others, like phosphorus, to form a compound whose solution changes blue litmus-paper to red; and with others, like hydrogen, to form a compound which has no effect on either the red or the blue paper.

The first class of compounds are called *bases;* the second class, *acids;* and the third class, *neutrals.*

85. *The Names of Acids.*—The acid just formed by the combination of oxygen and phosphorus is called *phosphoric* acid. The acid compounds of oxygen are always named from the element with which the oxygen combines, as *phosphoric* acid from *phosphorus.*

We have seen in studying the compounds of nitrogen and oxygen, that oxygen often combines with the same element to form more than one acid. In such cases, the acids are distinguished by terminations or by prefixes. The termination *ous* indicates less oxygen than *ic.* Thus *nitrous* acid contains less oxygen than *nitric* acid.

The prefix *hypo* indicates less oxygen, and the prefix *hyper* more oxygen, than there is in the acid to whose name either is prefixed. Thus *hyponitric* acid contains less oxygen than nitric acid, and *hypochlorous* acid less oxygen than *chlorous* acid; while *hyperchloric* acid contains more oxygen than *chloric* acid.

86. *Names of Bases and Neutrals.*—The basic and neutral compounds of oxygen have the common name of *oxides.*

The bases are named from both elements. Thus *oxygen* and *sodium* combine to form a base called *oxide of sodium; oxygen* and *potassium,* to form the base *oxide of potassium.*

The compounds of nitrogen and oxygen also show that oxygen sometimes combines with the same element to form more than one oxide. In such cases, the name of the one which contains the more oxygen takes the ending *ic,* and the name of the other takes the ending *ous.* Thus $N_2\Theta$ contains less oxygen in proportion to the nitrogen than $N\Theta$, and the former is called *nitrous oxide,* the latter *nitric oxide.* So $FeΘ$ is called *ferrous oxide*, and $Fe_2Θ_3$ *ferric oxide;* $Hg_2Θ$ *mercurous oxide*, and $HgΘ$ *mercuric oxide.*

The compounds of nitrogen and oxygen also illustrate the fact, that when oxygen combines with the same element in several proportions, the higher compounds (that is, those containing the most oxygen) are likely to be *acids*, while the lower are *oxides* (bases or neutrals).

87. *Oxyacids and Hydracids.* — In the case of muriatic acid, we have, as we have seen, an acid made up of hydrogen and chlorine, and containing no oxygen. There are several other acids which contain hydrogen as a common element instead of oxygen.

The acids whose common element is oxygen are called *oxyacids*, and those whose common element is hydrogen are called *hydracids*.

88. *Names and Symbols of the Leading Hydracids.* — Hydrogen combines with all the non-metallic elements of the monad group, and with sulphur, to form *hydracids*. The symbols and names of these acids are as follows: —

HCl, hydrochloric acid (muriatic acid);
HF, hydrofluoric "
HBr, hydrobromic "
HI, hydriodic "
H_2S, hydrosulphuric " (sulphide of hydrogen, sulphuretted hydrogen).

It will be noticed that the *hydracids* are named from *both* their elements.

89. *Names and Symbols of the Leading Oxyacids.* — Oxygen often combines with a metal to form one or more acids, but the most important oxyacids are compounds of oxygen and a non-metallic element. The following are the most important: —

$N_2Θ_3$, nitrous acid;
$N_2Θ_5$, nitric "

$P_2\Theta_3$, phosphorous acid;
$P_2\Theta_5$, phosphoric "
$Cl_2\Theta_3$, chlorous "
$Cl_2\Theta_5$, chloric "
$As_2\Theta_3$, arsenious "
$As_2\Theta_5$, arsenic "
$B_2\Theta_3$, boracic "
$S\Theta_2$, sulphurous "
$S\Theta_3$, sulphuric "
$C\Theta_2$, carbonic "
$Si\Theta_2$, silicic "
$Cr\Theta_3$, chromic "

90. *The Bases of the Four Groups.* — The ordinary bases are compounds of oxygen and a metal. As oxygen is a *diatomic* element, the bases of the metals of the first group will regularly contain two atoms of the metal to one of oxygen. Their symbols and names are as follows: —

$Ag_2\Theta$, oxide of silver;
$K_2\Theta$, " potassium (potassa);
$Na_2\Theta$, " sodium (soda).

The bases of the metals of the *dyad* group will regularly contain one atom of the metal to one of oxygen. Their symbols and names are as follows: —

$Ba\Theta$, oxide of barium (baryta);
$Ca\Theta$, " calcium (lime);
$Cd\Theta$, " cadmium;
$Co\Theta$, " cobalt;
$Cu\Theta$, " copper;
$Mg\Theta$, " magnesium (magnesia);
$Mn\Theta$, " manganese;
$Ni\Theta$, " nickel;
$Pb\Theta$, " lead;
$Sr\Theta$, " strontium (strontia);
$Zn\Theta$, " zinc;

Hg_2O, mercurous oxide;
HgO, mercuric "
FeO, ferrous "
Fe_2O_3, ferric "
CrO, chromous "
Cr_2O_3, chromic "
UO, uranous "
U_2O_3, uranic "

It will be noticed that mercury, besides its regular base, forms another base analogous to those of the metals of the monad group. Also, that iron, chromium, and uranium, besides their regular bases, form others analogous to those of the metals of the triad group.

The bases of the *triad* group regularly contain two atoms of the metal to three of oxygen. They are the following: —

Al_2O_3, oxide of aluminium (alumina);
Sb_2O_3, " antimony;
Bi_2O_3, " bismuth;
Au_2O_3, " gold.

Each of the metals of the *tetrad* group forms two oxides:—

SnO, stannous oxide;
SnO_2, stannic "
PtO, platinous "
PtO_2 platinic "

SULPHIDES AND CHLORIDES OF THE METALS OF THE FOUR GROUPS.

91. *Sulphides.* — Sulphur combines with all the metals of the four groups. As sulphur is a *diatomic* element, the sulphides have the same composition as the oxides. The metals which have two oxides have usually two corresponding sulphides.

92. *Chlorides.* — As chlorine is a *monatomic* element, the chlorides of the metals of the *monad* group regularly contain one atom of the metal to one of chlorine. Those of the metals of the *dyad* group will contain one atom of the metal to two of chlorine; of the *triad* group, one of the metal to three of chlorine; and of the *tetrad* group, one of the metal to four of chlorine.

The metals which have two oxides have usually two corresponding chlorides.

93. *Problems.* — 25. Write the symbols and names of the sulphides of the metals of the four groups.

26. Write the symbols and names of the chlorides of the metals of the four groups.

HYDRATES.

94. *Hydrates of the Acids.* — The oxyacids which have the composition represented above (89) are rare substances. These acids, with the exception of carbonic acid, ordinarily exist in combination with one or more molecules of water, and are then called *hydrates.* When not combined with water they are usually called *anhydrous* acids, or more briefly *anhydrides,* — names derived from the Greek, and meaning *without water.*

95. *Symbols of the Hydrates of the Acids.* — The symbols of these hydrates are formed by writing the symbol of water before the symbol of the anhydrous acid, with a comma between the two. Thus the symbol of ordinary sulphuric acid (oil of vitriol) is $H_2\Theta, S\Theta_3$, and that of ordinary nitric acid (aqua-fortis) is $H_2\Theta, N_2\Theta_5$. When more than one molecule of water combines with a molecule of the anhydride to form the hydrate, the number of the molecules of water is indicated by a figure placed before the symbol $H_2\Theta$. Thus the symbol for ordinary phosphoric acid, $3H_2\Theta, P_2\Theta_5$, indicates that three molecules of water

and one of *phosphoric anhydride* combine to form a molecule of the hydrate.

96. *Dualistic and Unitary Symbols.* — It is evident that the ultimate composition of the hydrate of sulphuric acid can be indicated by the symbol $H_2S\Theta_4$, since $H_2\Theta$, $S\Theta_3 = H_2S\Theta_4$. So the ultimate composition of hydrate of nitric acid can be represented by $HN\Theta_3$, since $H_2\Theta$, $N_2\Theta_5 = H_2N_2\Theta_6 = 2HN\Theta_3$; and that of the hydrate of phosphoric acid by $H_3P\Theta_4$, since $3H_2\Theta$, $P_2\Theta_5 = H_6P_2\Theta_8 = 2H_3P\Theta_4$.

The symbol $H_2\Theta$, $S\Theta_3$ represents a molecule of the hydrate of sulphuric acid as made up of a molecule of water and a molecule of sulphuric anhydride. But we may regard a molecule of this hydrate as merely containing the elements hydrogen, sulphur, and oxygen; and may represent it, therefore, by the symbol $H_2S\Theta_4$. The first method of writing the symbol is called the *dualistic* method; the second, the *unitary* method.

If we have two small piles of bricks we can make them into one, either by bringing them together without disarranging the bricks in either pile, or we can take the bricks one by one from each pile and arrange them in a new pile. So the atoms in a molecule of water and a molecule of sulphuric acid may be combined into a new molecule, either by bringing them together without disarranging the atoms which make them up, or by taking the atoms from each of the two molecules and arranging them into a new one. The *dualistic* symbol represents a molecule of the hydrate as made up in the first way; the *unitary*, in the second way.

Since the dualistic symbol indicates at a glance that the hydrate is formed by the union of water and the anhydride, it is often convenient to use it, although it may not correctly represent the molecular constitution of the hydrate.

The *hydracids* never form hydrates.

97. *Hydrates of the Bases.*—The bases also are usually

found in combination with water, and are then called *hydrates*. Thus $Na_2\Theta$, $H_2\Theta$ is the *hydrate of sodium;* $Ba\Theta$, $H_2\Theta$ is the *hydrate of barium;* and $Al_2\Theta_3$, $3H_2\Theta$ is the *hydrate of aluminium.*

98. *Symbols of these Hydrates.* — The symbol for the hydrate of a base is formed in the same way as that for the hydrate of an acid, except that the symbol of the water is put *after* that of the base.

These symbols also may be written according to the *unitary* method. Thus $HNa\Theta$ may represent the *hydrate of sodium,* since $Na_2\Theta$, $H_2\Theta = 2HNa\Theta$; $H_2Ba\Theta_2$ may represent the *hydrate of barium,* since $Ba\Theta$, $H_2\Theta = H_2Ba\Theta_2$; and $H_3Al\Theta_3$ the *hydrate of aluminium,* since $Al_2\Theta_3$, $3H_2\Theta = 2H_3Al\Theta_3$.

99. *Hydrate of Ammonia.* — Ammonia gas also combines with water to form a hydrate analogous to the hydrate of sodium. $H_3N + H_2\Theta = H_3N, H_2\Theta$. The resemblance of this hydrate to the hydrate of sodium is shown by writing the symbol thus: $H(H_4N)\Theta$. This symbol will of course correctly represent the composition of the hydrate, since $H_3N, H_2\Theta = H(H_4N)\Theta$. Comparing this formula with the formulas for hydrate of potassium and sodium, $HK\Theta$ and $HNa\Theta$, we see that the group of atoms (H_4N) takes the place of the atom of potassium in the one hydrate, and of the atom of sodium in the other.

100. *Compound Radicals.* — *Ammonium.* — There are many groups of atoms which seem in compounds to play the part of a single atom. Such groups of atoms are called *compound radicals.* Some of them are capable of existing by themselves, or in a free state; others are not. The group (H_4N) is called *ammonium.* It has never yet been obtained in a free state. In combining with other elements it always plays the part of a monad metal.

There is also a compound of nitrogen and carbon, which acts like a non-metallic monad element. It is called *cy-*

anogen and its symbol is ЄN, or Cy. It forms a hydracid, HЄN, or HCy, called *hydrocyanic acid.* Cyanogen can exist in a free state.

101. *Some Bases form more than one Hydrate.*—In many cases more than one hydrate is formed. Those given above are regarded as the regular or *normal* hydrates. The normal hydrates of the bases of the monad group have a composition corresponding to that of the hydrate of sodium; those of the dyad group, a composition corresponding to that of the hydrate of barium; and those of the triad group, a composition corresponding to that of the hydrate of aluminium.

102. *Problems.*—27. Write the *dualistic* and the *unitary* symbols of the hydrates of the bases of the *monad, dyad,* and *triad* group.

28. What fraction of the weight of hydrate of sulphuric acid is water? What fraction is sulphuric anhydride?

29. In the same hydrate, what fraction of the weight is hydrogen? sulphur? oxygen?

30. How much phosphoric anhydride can be obtained from 135 grammes of hydrate of phosphoric acid? How much phosphorus?

31. How much hydrate of sodium may be made from 3 kilogrammes of sodium? How much water will be required? How much oxygen?

SALTS.

103. If dilute nitric acid be added to a solution of hydrate of sodium, and the mixture be slowly evaporated, a white crystalline solid will be formed. This substance is found on examination to be wholly unlike either nitric acid or hydrate of sodium. Its solution has no effect on either red or blue litmus-paper. If a platinum wire be dipped into the solution and held in the flame of an alcohol or Bunsen's lamp, the intense yellow color of the flame is at

once recognized as that which sodium gives to a flame (46). Hence this compound must contain sodium.

If a solution of ferrous sulphate (copperas, or green vitriol) be added to dilute nitric acid, the mixture will at once become dark brown. This color will pass off on heating the liquid.

If we add strong sulphuric acid to the solution of the crystalline substance obtained above, and then add some ferrous sulphate, the mixture becomes dark brown; but on heating it, this color passes off. This substance then contains nitric acid.

On careful analysis the substance is found to contain sodium, oxygen, and nitrogen in the proportions indicated by the formula $NaN\Theta_3$.

What changes, then, took place on mixing the nitric acid and the hydrate of sodium? They can be best shown by the following equation:

$$H_2\Theta, N_2\Theta_5 + Na_2\Theta, H_2\Theta = Na_2\Theta, N_2O_5 + 2H_2\Theta;$$

or, using the unitary symbols:

$$HN\Theta_3 + HNa\Theta = NaN\Theta_3 + H_2\Theta.$$

In these equations it is seen that the hydrogen of the nitric acid has changed places with the sodium of the hydrate of sodium.

The first equation shows at a glance that this compound contains the elements of oxide of sodium and nitric anhydride. From its composition it is called *nitrate of sodium.*

A substance which contains the elements of an anhydrous base and acid is called a *salt.*

104. *Symbols of Salts.* — The *dualistic* symbol for a salt is obtained by writing the symbol of the acid after the symbol of the base, with a comma between them. Thus the dualistic symbol of *nitrate of sodium* is $Na_2\Theta$, $N_2\Theta_5$. The unitary symbol is obtained by writing the symbols of the elements in succession, placing the symbol of the metal

first, and that of oxygen last; as, for *nitrate of sodium*, $NaN\Theta_3$.

The dualistic symbol represents a molecule of the salt as made up by the combination of one or more molecules of a base with one or more molecules of an acid. The unitary symbol represents the salt as made up of three elements, without reference to the arrangement of the atoms in the molecule.

The dualistic symbol indicates at once that a salt is made up of the elements of a base and an acid, and the well-known fact that the salt can be again separated into a base and acid. It also represents the exact composition of the salt. But salts may be decomposed in other ways than into a base and an acid. Thus, when the electric current is sent through a solution of nitrate of sodium, $NaN\Theta_3$, it is separated into sodium, or Na, and $N\Theta_3$. To represent this and similar decompositions of the salt, the unitary symbol is the better. Both symbols will be used in this book according to convenience.

105. *Names of Salts.* — The name of the salt is formed by changing the ending *ic* of the name of the acid into *ate*, or the ending *ous* into *ite*, and adding the name of the metal of the base. Thus *nitric acid* and *oxide of sodium* form *nitrate of sodium*, and *nitrous acid* and *oxide of potassium* would form *nitrite of potassium.*

106. *The Hydrates are really Salts.* — It will be seen from the above formulas that the hydrates of the acids are salts in which water plays the part of the base, and that the hydrates of the bases are salts in which water plays the part of the acid.

107. *The Action of the Hydrates of the Acids on the Hydrates of the Bases.* — When the hydrates of the acids act on the hydrates of the bases, the metal of the base always changes place with the hydrogen of the acid, giving rise to a salt and water. As these salts contain three elements, they are called *ternary* salts.

The following equations represent the action of the hydrates of nitric and sulphuric acid upon the hydrate of a base in each of the first three groups :—

(1.) $K_2\Theta, H_2\Theta + H_2\Theta, N_2\Theta_5 = 2H_2\Theta + K_2\Theta, N_2\Theta_5$,

or $HK\Theta + HN\Theta_3 = H_2\Theta + KN\Theta_3$;

$Na_2\Theta, H_2\Theta + H_2\Theta, S\Theta_3 = 2H_2\Theta + Na_2\Theta, S\Theta_3$,

or $2HNa\Theta + H_2S\Theta_4 = 2H_2\Theta + Na_2S\Theta_4$.

(2.) $Ba\Theta, H_2\Theta + H_2\Theta, N_2\Theta_5 = 2H_2\Theta + Ba\Theta, N_2\Theta_5$,

or $H_2Ba\Theta_2 + 2HN\Theta_3 = 2H_2\Theta + BaN_2\Theta_6$;

$Ba\Theta, H_2\Theta + H_2\Theta, S\Theta_3 = 2H_2\Theta + Ba\Theta, S\Theta_3$,

or $H_2Ba\Theta_2 + H_2S\Theta_4 = 2H_2\Theta + BaS\Theta_4$.

The symbol for *nitrate of barium* is very often written $Ba(N\Theta_3)_2$, or $Ba2N\Theta_3$, instead of $BaN_2\Theta_6$. We have seen that the atom of potassium has the same atom-fixing power as an atom of hydrogen. And here the group of atoms $N\Theta_3$ has the same atom-fixing power as an atom of hydrogen, since they combine with one, and only one, atom of potassium. But an atom of barium has the same atom-fixing power as two atoms of hydrogen ; hence it requires double the group of atoms $N\Theta_3$ to satisfy it. This fact is at once indicated by the symbol $Ba(N\Theta_3)_2$ or $Ba2N\Theta_3$. The group of atoms $S\Theta_4$, on the other hand, has the same atom-fixing power as two atoms of hydrogen ; hence it satisfies two atoms of sodium, but only one of barium.

(3.) $Al_2\Theta_3, 3H_2\Theta + 3(H_2\Theta, N_2\Theta_5) = 6H_2\Theta + Al_2\Theta_3, 3N_2\Theta_5$,

or $H_3Al\Theta_3 + 3HN\Theta_3 = 3H_2\Theta + AlN_3\Theta_9$ or $Al(N\Theta_3)_3$;

$Al_2\Theta_3, 3H_2\Theta + 3(H_2\Theta, S\Theta_3) = 6\ H_2\Theta + Al_2\Theta_3, 3S\Theta_3$,

or $2H_3Al\Theta_3 + 3H_2S\Theta_4 = 6H_2\Theta + Al_2S_3\Theta_{12}$ or $Al_2(S\Theta_4)_3$.

The symbols and names of the ternary salts obtained in the above equations are as follows:—

Dualistic Symbol.	Unitary Symbol.	Name.
$K_2\Theta, N_2\Theta_5$	$KN\Theta_3$	Nitrate of potassium
$Ba\Theta, N_2\Theta_5$	$Ba(N\Theta_3)_2$	" barium
$Al_2\Theta_3, 3N_2\Theta_5$	$Al(N\Theta_3)_3$	" aluminium
$Na_2\Theta, S\Theta_3$	$Na_2S\Theta_4$	Sulphate of sodium
$Ba\Theta, S\Theta_3$	$BaS\Theta_4$	" barium
$Al_2\Theta_3, 3S\Theta_3$	$Al_2(S\Theta_4)_3$	" aluminium

The atom-fixing power of the groups of atoms $S\Theta_4$ and $N\Theta_3$ is strikingly shown by the following table:—

$K_2\Theta$	$K_2S\Theta_4$	KCl	$KN\Theta_3$
$Ba\Theta$	$BaS\Theta_4$	$BaCl_2$	$Ba(N\Theta_3)_2$
$Al_2\Theta_3$	$Al_2(S\Theta_4)_3$	$AlCl_3$	$Al(N\Theta_3)_3$

Neither of the groups of atoms $S\Theta_4$ and $N\Theta_3$ can exist in an uncombined state. The above formulas are not intended to convey the idea that $S\Theta_4$ and $N\Theta_3$ exist in the various salts as distinct compounds, but merely to express the fact that the group of atoms represented by $S\Theta_4$ is equivalent in combination to one atom of oxygen or two of hydrogen, and the group of atoms represented by $N\Theta_3$ is equivalent in combination to one atom of hydrogen.

108. *Normal, Acid, and Basic Salts.*—When the hydrate of sulphuric acid acts upon the oxide of sodium, half of the hydrogen of the acid may be displaced by the sodium. This change is best represented by unitary symbols:

$$Na_2\Theta + H_2S\Theta_4 = HNa\Theta + HNaS\Theta_4.$$

This compound is regarded as a salt, and is called an *acid* salt, or *super*-salt; while the salt $Na_2S\Theta_4$, in which all the hydrogen of the acid has been replaced by sodium, is called a *normal* salt.

A normal salt, then, is one in which all the hydrogen of the acid has been replaced by an equivalent of a metal. It will be seen by the above formula for acid sulphate of sodium, that it contains more oxygen in proportion to the sodium than the normal salt $Na_2S\Theta_4$. Hence an acid salt may be defined as one which contains more oxygen in proportion to the metal than a normal salt.

Salts sometimes contain more of the metal in proportion to the oxygen than a normal salt, and in that case are called *basic salts*, or *sub-salts.*

109. *Monobasic, Dibasic, and Tribasic Acids.* — An acid, like the nitric, which can form only one salt with potassa or soda, is called *monobasic.* An acid, like the sulphuric, which can form two salts with potassa or soda, is called *dibasic*, or *bibasic.* An acid which can form three salts with the base is called *tribasic.* Phosphoric acid is tribasic. The hydrate of the acid and the three salts which the acid forms with oxide of sodium are represented by the following formulas:—

$H_3P\Theta_4$, phosphoric acid;
$H_2NaP\Theta_4$, monobasic phosphate of sodium;
$HNa_2P\Theta_4$, dibasic " "
$Na_3P\Theta_4$, tribasic " "

or

$3H_2\Theta, P_2\Theta_5$, phosphoric acid;
$Na_2\Theta, 2H_2\Theta, P_2\Theta_5$, monobasic phosphate of sodium;
$2Na_2\Theta, H_2\Theta, P_2\Theta_5$, dibasic " "
$3Na_2\Theta, P_2\Theta_5$, tribasic " "

110. *Problems.* — 32. Show the action of the hydrates of sulphuric and nitric acid on the hydrates of the bases of the monad, dyad, and triad groups, using both the dualistic and unitary formulas.

33. What is the name of each of the ternary salts formed?

34. In each of these salts, what fraction of the whole weight does each element form?

35. Show by the unitary formulas the action of the hydrates of sulphuric and phosphoric acids on the oxides of potassium, sodium, and ammonium (100), giving all the salts that may be formed.

☞ In these problems it is understood that *normal* salts are to be obtained, unless otherwise stated.

TERNARY SALTS FORMED BY SUBSTITUTION.

111. Ternary salts are formed, not only by the action of the hydrates of the acids upon the hydrates of the bases, but also by the action (1.) of the hydrates of bases upon ternary salts, (2.) of the hydrates of acids upon ternary salts, and (3.) of ternary salts upon each other.

1. Thus when a solution of *hydrate of sodium* is added to a solution of *nitrate of lead*, the sodium and the lead change places, and *hydrate of lead* and *nitrate of sodium* are formed.

$$Na_2\Theta, H_2\Theta + Pb\Theta, N_2\Theta_5 = Pb\Theta, H_2\Theta + Na_2\Theta, N_2\Theta_5;$$
$$\text{or,}\quad 2HNa\Theta + Pb(N\Theta_3)_2 = H_2Pb\Theta_2 + 2NaN\Theta_3.$$

The hydrate of lead is insoluble, and separates as a solid, while the nitrate of sodium remains in solution. When a solid separates in this way from a solution, it is called a *precipitate.*

2. When *hydrate of sulphuric acid* is added to a solution of *nitrate of lead*, the hydrogen and the lead change places, forming *sulphate of lead*, which is insoluble, and *hydrate of nitric acid.*

$$H_2\Theta, S\Theta_3 + Pb\Theta, N_2\Theta_5 = Pb\Theta, S\Theta_3 + H_2\Theta, N_2\Theta_5;$$
$$\text{or,}\quad H_2S\Theta_4 + Pb(N\Theta_3)_2 = PbS\Theta_4 + 2HN\Theta_3.$$

3. When a solution of *sulphate of calcium* is added to a solution of *nitrate of barium*, the calcium and barium change places, forming *sulphate of barium* and *nitrate of calcium.*

$$Ca\Theta, S\Theta_3 + Ba\Theta, N_2\Theta_5 = Ba\Theta, S\Theta_3 + Ca\Theta, N_2\Theta_5;$$
$$\text{or,}\quad CaS\Theta_4 + Ba(N\Theta_3)_2 = BaS\Theta_4 + Ca(N\Theta_3)_2.$$

It appears in each of the above equations that the new compounds are formed by *substitution.*

DOUBLE DECOMPOSITION.

112. When sodium is thrown upon water, an atom of sodium, as we have seen (52), replaces an atom of hydrogen in the molecule of water, giving rise to a molecule of hydrate of sodium.

$$H_2\Theta + Na = HNa\Theta + H.$$

Here, as only one substance is decomposed, the change which takes place may be described as *substitution* by *single decomposition.* In the cases given above (111), it will be noticed that both substances are decomposed; hence the change is described as substitution by *double decomposition.*

Experiments have brought to light the following important law:—

When the solutions of two compounds are mixed, decomposition always takes place when thereby an insoluble or gaseous compound can be formed.

In such cases, 1 atom of a *monad* is regularly replaced by 1 atom of a monad; 1 atom of a *dyad* by 1 atom of a dyad or 2 of a monad; 1 atom of a *triad* by 1 of a triad or 3 of a monad, or 2 atoms of a triad by 3 of a dyad; and 1 atom of a *tetrad* by 1 of a tetrad, 4 of a monad, or 2 of a dyad, or 3 atoms of a tetrad by 4 of a triad.

113. *Affinity modified by Cohesion.* — Few solid compounds will act upon each other when brought together, while many of them will readily act upon each other *in solution.* The cohesion of the solid acts against the affinity between the elements of the two substances, preventing them from entering into combination. When the solids are dissolved this cohesion is overcome.

BINARY SALTS.

114. The action of hydrochloric acid upon hydrate of sodium is shown in the following equation:

$$Na_2\Theta, H_2\Theta + 2HCl = 2NaCl + 2H_2\Theta;$$

or,

$$HNa\Theta + HCl = NaCl + H_2\Theta.$$

When a hydracid acts upon the hydrate of a base, a *binary* compound is formed. This binary compound, since it results from the action of an acid upon a base, is called a *binary salt.*

115. *Names and Symbols of Binary Salts.* — The name of a binary salt is formed by changing the ending of the name of the non-metallic element into *ide* and adding the name of the metal. Thus the salt obtained above is called *chloride of sodium.* The salt obtained by the action of *hydriodic acid* on *hydrate of potassium* is called *iodide of potassium. Hydrocyanic* acid (100) forms binary salts called *cyanides.*

The symbol of a binary salt is formed by writing the symbol of the non-metallic after the metallic element. The symbols for the two salts just mentioned are NaCl and KI.

ACTION OF HYDRACIDS UPON BASES OF THE MONAD, DYAD, AND TRIAD GROUPS.

116. — 1. The action of hydrochloric and hydrosulphuric acids upon hydrate of potassium is represented by the following equations:

$$HK\Theta + HCl = KCl + H_2\Theta;$$
$$2HK\Theta + H_2S = K_2S + 2H_2\Theta.$$

The action of these acids upon other bases of the monad group is precisely like their action on hydrate of potassium.

2. The action of the same acids on hydrate of barium is as follows:

$$H_2Ba\Theta_2 + 2HCl = BaCl_2 + 2H_2\Theta.$$
$$H_2Ba\Theta_2 + H_2S = BaS + 2H_2\Theta.$$

Their action upon the other regular bases of the dyad group is the same.

3. Their action upon hydrate of aluminium is as follows:

$$H_3Al\Theta_3 + 3HCl = AlCl_3 + 3H_2\Theta.$$
$$2H_3Al\Theta_3 + 3H_2S = Al_2S_3 + 6H_2\Theta.$$

Their action on other triad bases is the same. The action of hydrobromic, hydrocyanic (100), hydrofluoric, and hydriodic acids upon bases is like that of hydrochloric acid.

It will be noticed in the first of the above equations that only one molecule of the base $HK\Theta$ and one of the acid HCl are represented, since only one atom of potassium and one of chlorine are needed to form a molecule of chloride of potassium, KCl. When hydrosulphuric acid, H_2S, is represented as acting on the same base, $HK\Theta$, two molecules of the base are represented, since two atoms of potassium are required to combine with one atom of sulphur to form a molecule of sulphide of potassium, K_2S. Chloride of barium, $BaCl_2$, contains two atoms of chlorine; hence two molecules of hydrochloric acid are represented as acting upon one molecule of the base $H_2Ba\Theta_2$. But since sulphide of barium, BaS, contains but one atom of sulphur, only one molecule of hydrosulphuric acid, H_2S, is represented as acting upon the same base, $H_2Ba\Theta_2$. For a like reason, three molecules of hydrochloric acid, HCl, are represented as acting upon one of the base $H_3Al\Theta_3$, and three molecules of hydrosulphuric acid, H_2S, as acting upon two molecules of $H_3Al\Theta_3$.

117. *Problems.* — 36. Show the action of hydrochloric

and hydrosulphuric acids upon the bases of the monad, dyad, and triad groups, using the unitary formulas.

37. How much metal in 50 kilogrammes of each binary salt formed?

38. How much iodine in a kilogramme of iodide of potassium? How much chlorine in the same amount of chloride of potassium?

BINARY SALTS FORMED BY SUBSTITUTION.

118. Binary salts are also formed by the action (1.) of hydracids upon binary salts, (2.) of binary salts upon each other, (3.) of hydracids upon ternary salts, and (4.) of binary salts upon ternary salts. The following equations will serve to illustrate the different cases:

(1.) $2AsCl_3 + 3H_2S = As_2S_3 + 6HCl$;

(2.) $BiCl_3 + 3KI = 3KCl + BiI_3$;

(3.) $Pb(N\Theta_3)_2 + 2HCl = 2HN\Theta_3 + PbCl_2$;

(4.) $2AgN\Theta_3 + \text{Ȼ}aCl_2 = \text{Ȼ}a(N\Theta_3)_2 + 2AgCl$.

THE LAW OF DOUBLE DECOMPOSITION.

119. In the hydrates of acids it is evident that hydrogen plays the part which the metal does in a ternary salt, while in hydracids it plays the part which the metal does in a binary salt. For these and other reasons hydrogen has been regarded by many chemists as a *metal.* If hydrogen in the two cases just named be regarded as a metal, and if, consequently, the hydrates of the acids and bases be regarded as ternary salts, and the hydracids as binary salts, all the cases of double decomposition given above may be brought under this one general rule:—

When two salts on the mixture of their solutions undergo mutual decomposition, the metals change places.

As before stated, *the salts always undergo this mutual decomposition when an insoluble or volatile compound can thus be formed.*

120. *Problems.* — 39. Show the reaction between nitrate of lead and hydrate of sodium.

☞ Any chemical change is called a *reaction.*

40. Show the reaction between nitrate of silver and hydrate of ammonium.

41. Nitrate of silver and chromate of potassium.

42. Nitrate of lead and chromate of potassium.

43. Mercurous nitrate and chromate of potassium.

44. Nitrate of silver and hydrate of sulphuric acid.

45. Nitrate of lead and hydrate of sulphuric acid.

46. Sulphate of copper and hydrate of potassium.

47. Sulphate of copper and chromate of sodium.

48. Sulphate of manganese and hydrate of sodium.

49. Sulphate of zinc and hydrate of potassium.

50. Nitrate of cobalt and hydrate of sodium.

51. Sulphate of nickel and hydrate of potassium.

52. Sulphate of manganese and carbonate of ammonium.

53. Nitrate of barium and carbonate of ammonium.

54. Sulphate of calcium and carbonate of ammonium.

55. Nitrate of strontium and sulphate of calcium.

56. Nitrate of barium and chromate of sodium.

57. Sulphate of aluminium and hydrate of ammonium.

58. Hydrate of ammonium and sulphate of chromium.

59. Ferrous nitrate and hydrate of ammonium.

60. Hydrochloric acid and nitrate of silver.

61. Hydrochloric acid and nitrate of lead.

62. Hydrochloric acid and mercurous nitrate.

63. Hydrosulphuric acid and nitrate of silver.

64. Hydrosulphuric acid and nitrate of lead.

65. Hydrosulphuric acid and mercurous nitrate.

66. Nitrate of silver and iodide of potassium.

67. Iodide of potassium and nitrate of lead.

68. Iodide of potassium and mercurous nitrate.
69. Stannous chloride ($SnCl_2$) and hydrosulphuric acid.
70. Stannic chloride ($SnCl_4$) and hydrosulphuric acid.
71. Chloride of bismuth and hydrosulphuric acid.
72. Stannous chloride and hydrate of potassium.
73. Chloride of gold and hydrosulphuric acid.
74. Chloride of platinum and iodide of potassium.
75. Chloride of gold and iodide of potassium.
76. Mercuric chloride and sulphide of hydrogen.
77. Chloride of bismuth and sulphuretted hydrogen.
78. Chloride of cadmium and hydrosulphuric acid.
79. Sulphate of copper and hydrosulphuric acid.
80. Chloride of cadmium and hydrate of potassium.
81. Mercuric chloride and iodide of potassium.
82. Sulphate of copper and iodide of potassium.
83. Chloride of cadmium and chromate of potassium.
84. Chloride of aluminium and hydrate of ammonium.
85. Chloride of barium and carbonate of ammonium.
86. Chloride of strontium and carbonate of ammonium.
87. Chloride of calcium and carbonate of ammonium.
88. Chloride of strontium and sulphate of calcium.
89. Chloride of barium and hydrate of sodium.
90. Chloride of barium and carbonate of sodium.
91. Chloride of barium and silicate of potassium.
92. Chloride of calcium and silicate of potassium.
93. Chloride of calcium and carbonate of sodium.
94. Nitrate of silver and carbonate of potassium.
95. Nitrate of silver and silicate of potassium.
96. Chloride of barium and fluoride of potassium.
97. Chloride of calcium and fluoride of sodium.
98. Nitrate of silver and chloride of sodium.
99. Nitrate of silver and bromide of potassium.*

* See Appendix, 28.

SUMMARY.

It is possible to decompose water into hydrogen and oxygen; but there is no way known by which hydrogen and oxygen can be decomposed into simpler substances. We therefore divide substances into two classes:—

1st. *Elements*, or substances which cannot be decomposed by any known process. Of these only 65 are known.

2d. *Compounds*, or substances made up of elements. (46–49.)

The *force* which causes elements to combine is called *Affinity*.

Water is wholly unlike either hydrogen or oxygen, or the mere mechanical mixture of the two gases. This illustrates the first characteristic of affinity; that *it changes the properties of the substances which it causes to combine.*

Hydrogen and oxygen always exist in water in the proportion of two parts of hydrogen *by volume* to one part of oxygen, or two parts of hydrogen *by weight* to sixteen parts of oxygen. This illustrates the second characteristic of affinity; that *it always causes substances to combine in fixed and definite quantities.* This law of combination is called the *Law of Definite Proportions.* (50.)

Sodium and potassium in acting upon water set the hydrogen free from the oxygen, and take its place in combination with the oxygen. This illustrates the third characteristic of affinity; that *a given element combines with some elements in preference to others.** (51.)

Potassium or sodium can displace half of the hydrogen from water, or the whole, giving rise to two compounds; the first containing potassium (or sodium), hydrogen, and oxygen; the second, potassium (or sodium) and oxygen.

* See Appendix, 20.

But there is no known substance which will displace any other fraction of hydrogen from a molecule of water, or any fraction of the oxygen without displacing the whole. Hence a molecule of water is made up of three parts which are *indivisible* by affinity, and which are hence called *atoms*. A molecule of water is made up of two atoms of hydrogen and one of oxygen. (52.)

Matter is, then, divisible in a threefold way: —

1st. By mechanical means into minute but sensible *masses*.

2d. By means of heat into insensible portions, called *molecules*.

3d. By means of affinity into *atoms*. (53.)

If the weight of the hydrogen atom be taken as unity, the weight of the oxygen atom will be 16. Since 23 parts by weight of sodium and 39 of potassium are required to displace one part of hydrogen, the atomic weights of those elements are 23 and 39 respectively. (52.)

The composition of substances and the chemical changes which they undergo can be briefly indicated by certain signs called *symbols*. (54.)

The symbol of an *element* is the first letter of its name, a second letter being added to distinguish names beginning with the same letter. The symbol of an element always stands for *one atom* of the element. (55.)

The symbol of a *compound* indicates its composition. It is formed by writing together the symbols of the elements of the compound, with a small figure after each symbol expressing the number of atoms of that element found in a molecule of the compound. The symbol of a compound always stands for *one molecule* of the compound. (56.)

The chemical changes, or *reactions*, which substances undergo, are indicated by *equations* made up of these symbols. (57.)

On examining the three compounds, *muriatic acid*, *am-*

monia, and *marsh gas*, it appears that a molecule of the first is made up of *one* atom of hydrogen and one of chlorine; that a molecule of the second is made up of *three* atoms of hydrogen and one of nitrogen; and that a molecule of the third is made up of *four* atoms of hydrogen and one of carbon. (59 – 74.)

The atoms of different elements, then, have power to fix in combination a different number of atoms of hydrogen. Thus,

1	atom of	chlorine	can fix	1 atom	of hydrogen;
1	"	oxygen	"	2 atoms	"
1	"	nitrogen	"	3 "	"
1	"	carbon	"	4 "	"

Hence we classify the elements into four groups:—

1st. *Monatomic* elements, or *Monads;* those whose atom has the same atom-fixing power as *one* atom of hydrogen.

2d. *Diatomic* elements, or *Dyads;* those whose atom has the same atom-fixing power as *two* atoms of hydrogen.

3d. *Triatomic* elements, or *Triads;* those whose atom has the same atom-fixing power as *three* atoms of hydrogen.

4th. *Tetratomic* elements, or *Tetrads;* those whose atom has the same atom-fixing power as *four* atoms of hydrogen. (76 – 79.)

The compounds of nitrogen and oxygen show that the same elements may combine in more than one proportion, and that in such cases the proportions of the elements in the compounds are always *multiples* of the *atomic weight* of the elements. This law of combination is known as the *Law of Multiple Proportions.* (80 – 82.)

On examining the compounds of oxygen it is found that they can be arranged in three classes:—

1st. *Acids;* which in solution have the power to turn blue litmus-paper to red.

2d. *Bases;* which in solution turn red litmus-paper to blue.

3d. *Neutrals;* whose solutions have no effect on either red or blue litmus-paper. (84.)

Of *Acids* there are two classes:—

1st. *Oxyacids* (or *oxygen acids*), whose common element is *oxygen.*

2d. *Hydracids* (or *hydrogen acids*), whose common element is *hydrogen.* (87.)

Oxyacids are named from the element with which the oxygen combines; *hydracids* from *both* elements. (88, 89.)

Hydrogen never forms more than one acid with the same element. Oxygen often forms two or more acids with the same element.

When oxygen combines with the same element in several proportions, the *higher* compounds are likely to be *acids,* while the *lower* are either *bases* or *neutrals.* (86.)

When oxygen forms more than one acid with the same element, they are distinguished by the endings *ic* and *ous,* and by the prefixes *hypo* and *hyper,* and in some cases by other prefixes. (85.)

Bases and *neutrals* are called *oxides.* When two oxides are formed with the same element, they are distinguished by the endings *ic* and *ous,* and often by means of prefixes. (86.)

The oxyacids and the bases are usually combined with water, and are then called *hydrates.* (94.)

The composition of these hydrates can be represented by means of symbols in two ways:—

1st. By writing together the symbol of the *anhydrous acid* or *base* and the symbol of *water,* separating the two by a comma.

2d. By writing together the symbols of the *elements* contained in the hydrate.

The first method is called the *dualistic* method, since it represents the hydrate as formed by the combination of *two compounds.*

The second is called the *unitary* method, since it represents the hydrate as a compound containing three elements. (95 – 99.)

The hydrogen of the hydrate of an acid may be replaced by a metal, giving rise to a compound called a *salt.*

The composition of a salt, like that of a hydrate, may be represented by the dualistic and the unitary methods. The former represents the salt as made up of an *anhydrous base* and an *acid;* the latter represents it as a compound of *three elements.* (103, 104.) The salts which contain three elements are called *ternary* salts. (107.)

Ternary salts may be classified as follows:—

1st. *Normal salts;* those in which the hydrogen of the hydrate of the acid is wholly replaced by an equivalent of the metal.

2d. *Acid* salts; those which contain *less* metal in proportion to the oxygen than a normal salt.

3d. *Basic* salts; those which contain *more* metal in proportion to the oxygen than a normal salt. (108.)

Ternary salts are named from the acid and base which combine to form them; by changing the ending *ic* of the name of the acid into *ate*, and *ous* into *ite*, and adding the name of the metal of the base. (105.)

Certain groups of elementary atoms act in compounds like single elementary atoms. Such groups of atoms are called *compound radicals.* Some compound radicals can exist in a free state, others cannot. (100.)

When the hydrogen of the *hydracids* is replaced by a metal, a compound containing two elements is formed, which is called a *binary salt.* (114.)

Binary salts are named from the two elements which combine to form them; by changing the ending of the name of the non-metallic element into *ide*, and adding the name of the metal. (115.)

The *hydrates of the acids* may be regarded as *ternary salts* in which *hydrogen* plays the part of a *metal;* and the *hydracids* may be regarded as *binary salts*, in which hydrogen is the metal. (106, 107, 119.)

When the solutions of the salts are mixed, mutual decomposition takes place whenever an *insoluble* or *gaseous* compound can thus be formed. In this decomposition the metals of the salts merely change places; hydrogen being regarded as a metal. Hence the decomposition is said to take place by *substitution;* and, since both the original salts are decomposed, it is called *double decomposition.* (111, 112, 119.)

COMBUSTION AND ITS PRODUCTS.

121. When a lighted taper is held to a jet of ordinary coal-gas, the latter takes fire. In what does the burning of the coal-gas consist?

122. *The Products of the Burning of Coal-Gas are Carbonic Acid and Water.* — We will first find what is produced when coal-gas burns.

Invert a bottle over a gas-burner a short time; then remove it, pour in a little lime-water, close the mouth of the bottle with the hand, and shake it. The lime-water becomes milky-white. If lime-water is shaken in a bottle filled with air, it remains unchanged. Some new substance then has been formed by the burning of the gas.

Fasten a bit of charcoal to a wire, light it, and plunge it into a jar of oxygen. The charcoal glows brightly for a short time, and is then extinguished. If now we pour some lime-water into the jar and shake it, the liquid becomes milky-white; showing that the same substance is produced when carbon burns in oxygen as when gas burns in the air. This substance is of course a compound of carbon and oxygen, and since it will turn moistened blue litmus-paper to red, it is called *carbonic acid.* (84, 89.)

Carbonic acid, then, is one product of the burning of coal-gas.

If a jet of coal-gas be burned under a tin funnel connected with a long glass tube (which must be kept cold), moisture will soon collect on the inside of the tube. After a short time, enough will have collected to trickle down and drop from the end of the tube. If a bit of potassium be put into this liquid, it will burn with a rose-colored flame; showing that the liquid is water.

Fig. 33.

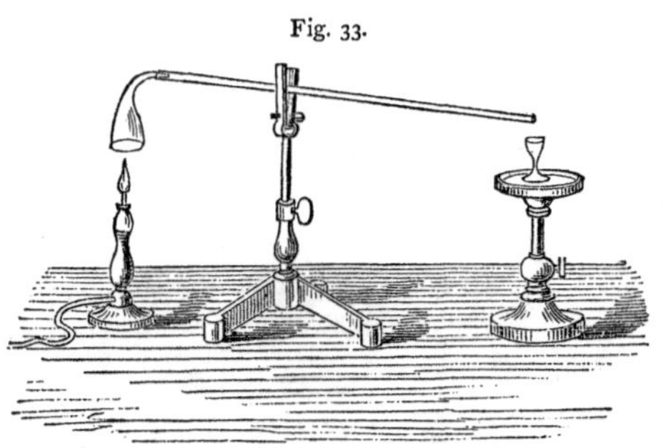

Water, then, is another product of the burning of coal-gas. It has been found that no other substance is produced, except in very minute quantities, by the burning of coal-gas.

123. *Does Coal-Gas in burning remove anything from the Air?* — Arrange a gas-burner so that it can be covered with a bell-jar whose mouth dips beneath the surface of water.* If now we light the gas and cover it with the jar, it soon ceases to burn, and the water rises in the jar. The coal-gas, then, in burning, does remove something from the air.

124. *The Coal-Gas removes Oxygen from the Air.* — Invert a jar of air over a jar of nitric oxide, and the gases on mixing become cherry-red, showing that there is oxygen in the air (80, 81).

Fig. 34.

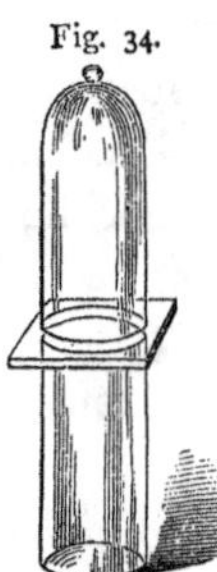

Burn a piece of phosphorus in a jar of air over water.† The jar is filled with dense white fumes, which are soon absorbed by the water, which rises and partially fills the jar. Introduce now some nitric oxide into the jar while still over the water, and the gas shows scarcely a trace of red color. It is evident,

* See Appendix, 21. † See Appendix, 22.

then, that the free oxygen has been almost wholly removed from the air by the burning phosphorus.

Introduce a jet of burning coal-gas into air from which the oxygen has been removed in this way, and it is at once extinguished.

Coal-gas in burning, then, removes oxygen from the air.

125. *The Burning of Coal-Gas consists in the Combination of the free Oxygen of the Air with the Carbon and Hydrogen of the Gas.* — Carbonic acid, one of the products of the burning of coal-gas, is, as we have learned (122), a compound of carbon and oxygen; water, the other product, we know to be a compound of hydrogen and oxygen. Of the three elements in these products, the oxygen, as we have seen, comes from the air; the carbon and hydrogen exist in the coal-gas.

The force, then, which causes the coal-gas to burn, is *affinity*, and the burning consists in the combination of the oxygen of the air with the elements of the gas.

126. *All ordinary Combustion consists in the Combination of the Oxygen of the Air with the burning Substance.* — We have already seen that carbon and phosphorus, as well as coal-gas, in burning combine with oxygen. If any ordinary combustible substance, such as a taper, wax, or wood, be ignited and plunged into a jar of air from which oxygen has been removed, it is instantly extinguished; showing that it cannot burn without a supply of oxygen. And if we burn any of these substances in a jar inverted over water, we see, as in the case of the phosphorus (124), that they remove something from the air. We therefore conclude that all ordinary combustion consists in the combination of the oxygen of the air with the burning body.

127. *Why a Draft is necessary in Stoves and Furnaces.* — Since combustion is the combination of oxygen with the burning body, we see why our stoves and furnaces must have a draft. As fast as the oxygen is taken from the air

by the burning fuel, this air must be removed, and a fresh supply must take its place. In other words, a stream of air must be kept constantly flowing over or through the fuel.

We see, also, how the fire can be regulated by means of the draft. If the doors or dampers through which the air is admitted be partially closed, the supply of air will be diminished, and the burning will therefore be retarded.

128. *Combustibles and Supporters of Combustion.* — Any substance, as coal-gas, which can be made to burn, is called a *combustible;* while any substance, as air or oxygen, in which it can burn, is called a *supporter of combustion.* These terms are convenient, though, strictly speaking, the one substance is no more a combustible or a supporter of combustion than the other. Since the burning of coal-gas consists in the combination of the gas with oxygen, the oxygen in reality burns, as well as the gas; and, on the other hand, the gas is as much a supporter of the combustion as the oxygen. The burning must of course take place where the gases come together. A jet of oxygen would appear to burn in an atmosphere of coal-gas, just as a jet of coal-gas appears to burn in an atmosphere of oxygen.

Fig. 35.

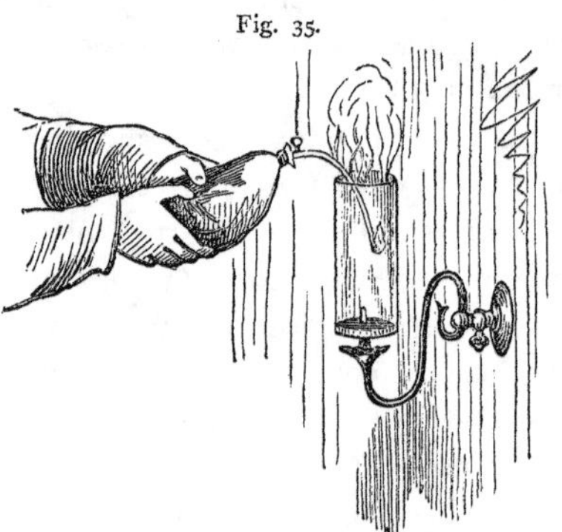

Fit a cork to one end of a lamp chimney, and let the tip of a gas-burner pass through it, as represented in the figure. Allow the gas to escape for some time, and then light it at the top of the chimney. It will burn quietly, and the chimney will evidently be filled with coal-gas. Fill a gas-bag with oxygen, and fasten to the bag a bent glass tube drawn out into a fine jet. Force the oxygen through the tube in a gentle stream, and introduce the end of the tube through the flame into the chimney. As it passes the flame, the oxygen takes fire and burns brightly in the coal-gas; the oxygen apparently becoming the combustible body, and the coal-gas the supporter of combustion.

In both the flames which we have here, it will be seen that gases are burning where they come together; — the coal-gas and the oxygen of the air, where they meet at the top of the chimney; the oxygen from the bag and the coal-gas, where they meet at the end of the tube.

129. *What Fraction of the Air is free Oxygen?* — When phosphorus is burned in a jar of air over water, it removes, as we have seen, the free oxygen from the air, and the water rises in the jar to take its place. After all the oxygen has combined with the phosphorus, and the compound formed has been absorbed by the water, the jar is found to be about one fifth filled with water; showing that about one fifth of the air is oxygen.

130. *Oxygen must be Heated before it will combine with ordinary Combustibles.* — The jet of coal-gas shows no disposition to burn until a lighted taper is applied to it. The oxygen of the air is at all times in contact with wood and coal, yet they do not burn unless they are first kindled. When even as inflammable a gas as hydrogen is mixed with oxygen, it does not burn unless ignited with a taper or an electric spark (50). At ordinary temperatures oxygen is one of the most passive of substances; but when heated it becomes very active. Its affinity is dormant until it is heated, when it is aroused to the intensest energy. When the lighted taper is held to the coal-gas, it heats the molecules of oxygen in contact with the gas, and rouses them to activity. These molecules then rush into combination with the gas with sufficient energy to develop the light and heat of the flame.

The passive condition of oxygen at ordinary temperatures, and the energy with which it rushes into combination when once aroused by heat, are shown by the following experiments.

Fill a rubber bag with a mixture of two measures of hydrogen and one of oxygen.* Attach a common clay pipe to the bag by a rubber tube, and blow some soap-bubbles with this mixture. Apply a lighted taper to these bubbles to heat the molecules of oxygen within. A violent explosion follows, showing with what energy the oxygen combines with the hydrogen when its affinity is once roused to activity.

Fig. 36.

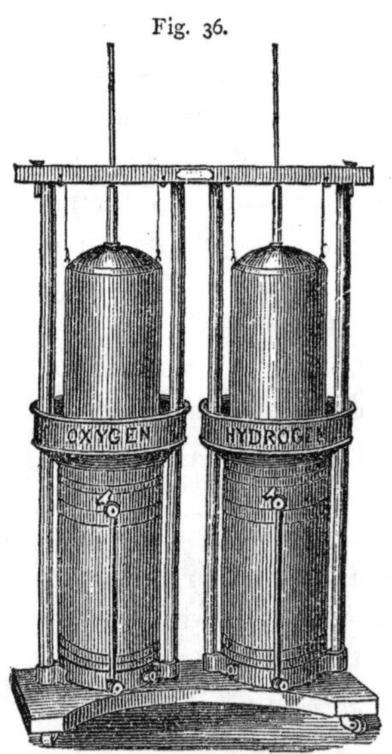

Fill one of the bell-jars of the gas-holder represented in the figure with oxygen, and the other with hydrogen. Connect the bell-jars with a burner by means of rubber tubes. This burner is so constructed that it allows the hydrogen to pass through it just twice as fast as the oxygen, and the gases can mix only at the end of the jet, as shown in Figure 37. Force the gases through the tubes by placing weights on the bell-jars, and ignite the mixture as it escapes. Hold a copper or iron wire in this flame, and it burns as readily as a pine shaving held in the flame of a lamp. A steel watch-spring burns with brilliant scintillations. If a bit of zinc is placed on a piece of charcoal hollowed out for the purpose, and this flame is directed upon it, the metal quick-

* See Appendix, 23.

ly melts and burns. Antimony, bismuth, and many other metals, will burn in the same way, each with a characteristic light. Cast-iron burns with a shower of bright sparks.

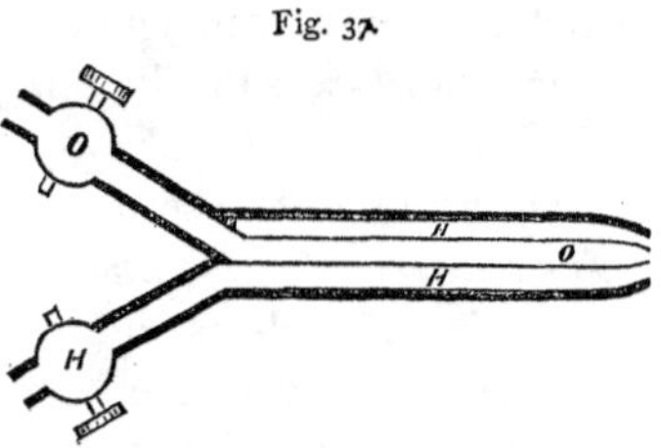

Fig. 37.

The flame produced in this way is called the *oxy-hydrogen* flame. Its intense heat, which is evident from the experiments just given, shows the energy with which the oxygen combines with the hydrogen.

131. *Combustion is Self-sustaining.* — It is not necessary to arouse any large amount of oxygen to activity in order to insure the continuance of the combustion. When, for instance, a lighted match is held to the wick of a candle, it excites but a few molecules of oxygen to activity. These few rush into combination with the elements of the candle, and by so doing develop sufficient heat to awaken the activity of more oxygen, which in turn enters into combination and develops more heat. In this way a supply of active oxygen is maintained until the candle is consumed.

132. *The Point of Ignition.* — Different substances begin to burn at very different temperatures. This is well illustrated in the kindling of a coal fire. Shavings are put into the grate first, then kindling-wood, then charcoal, and finally hard coal. The shavings are lighted by means of a match. The match is a bit of dry, soft wood, one end of which is covered with sulphur and tipped with phosphorus. It is a well-known fact, that when two bodies are rubbed together heat is developed. On striking the match sufficient heat is developed by the friction to ignite the phosphorus, which takes fire at a temperature of about 150° Fahrenheit. The

phosphorus in burning develops heat enough to ignite the sulphur, which burns at a temperature of about 500°. The burning sulphur develops heat enough to ignite the wood of the match; the match, to ignite the shaving; the shaving, the kindling-wood; the kindling-wood, the charcoal; and the charcoal, the hard coal, which requires the temperature of a full white heat to set it on fire.

133. *The Products of Combustion are not always Gaseous.* — In the burning of coal-gas and of a candle, the products are wholly gaseous, and in the burning of wood they are mainly gaseous; but when metals, as copper and iron, burn in the oxy-hydrogen flame, the products of their combustion are seen to be *solid.*

134. *The Burning of Metals in the Oxy-hydrogen Flame consists in their Combination with Oxygen.* — Bend a steel watch-spring into the form of a spiral, fasten a bit of wood to one end of it, light the wood, and plunge the spring into a jar of oxygen.* The steel takes fire, and burns with bright scintillations. The molten product of the combustion falls into the water in the bottom of the jar. If this product be carefully collected and weighed, it will be found to weigh more than the spring did at first. It is evident, then, that the iron in burning combines with oxygen.

When an iron wire is burned in the oxy-hydrogen flame, if all the products were collected and weighed, they would be found to weigh more than the wire; and since iron cannot be made to burn in hydrogen or in the air from which oxygen has been removed, we conclude that in burning in this flame it combines with oxygen.

If a piece of potassium or sodium be placed in a deflagrating spoon,† heated, and plunged into a jar of oxygen, it will burn brightly for a time. The product of the combustion is a white solid, which by the usual tests (84) we find to be *oxide of potassium* or *sodium.*

* See Appendix, 24. † See Appendix, 25.

135. *Magnesium and some other Metals will burn in the Air.* — A magnesium wire will burn brightly in the air. It may be lighted with a match. Calcium, also, burns readily in the air; as aluminium does if pulverized and strongly heated. Silicon, also, a non-metallic element, but somewhat like a metal in appearance, burns very readily in the air.

It will be noticed that all the elements just mentioned are rare in their free state. Many of the rare metals, as potassium, sodium, magnesium, and calcium, are as combustible as carbon; but they differ from carbon in giving rise to solid products while burning (133).

136. *Oxygen is not the only Supporter of Combustion.* — If a piece of gold-leaf or Dutch foil be dropped into a jar of chlorine, it disappears with a flash of light. Here the burning consists in a combination with chlorine. Tin and copper foil and pulverized antimony will also burn in chlorine. Many metals will burn in the vapor of sulphur; the burning being then a combination of the metal with sulphur.

137. *The Materials of the Earth's Crust are chiefly Chemical Compounds.* — If bits of marble are put into a bottle and hydrochloric acid poured over them, a violent effervescence takes place, and a gas is set free which may be col-

Fig. 38.

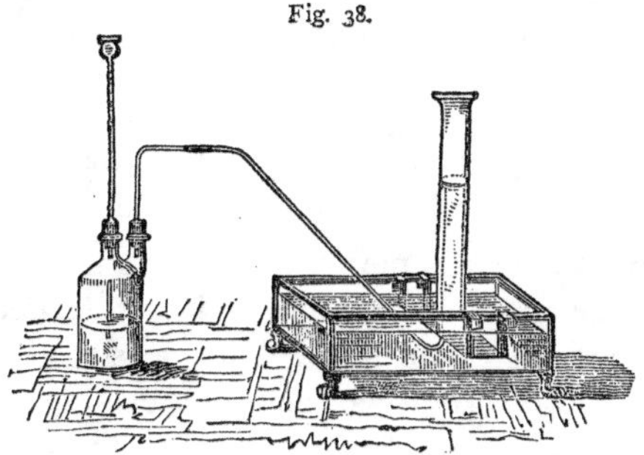

lected over water. If lime-water be poured into a jar of this gas and shaken, it turns milky-white, showing the gas to be carbonic acid (122). Marble, then, must contain carbonic acid, since hydrochloric acid contains only hydrogen and chlorine.

If the liquid which remains after the marble has been acted upon by the hydrochloric acid be evaporated to dryness, a white solid is obtained. This solid has been decomposed by the electric current into *chlorine* and a yellow metal called *calcium*; showing that it is a compound of chlorine and calcium. The chlorine evidently comes from the hydrochloric acid, and the calcium must have come from the marble. Marble, then, contains carbonic acid and calcium, and the chlorine of the hydrochloric acid in acting upon the marble sets the carbonic acid free. The existence of the metal calcium and of carbonic acid in marble seems to indicate that marble is *carbonate of calcium.* If so, its composition would be $CaCO_3$, and the action of hydrochloric acid upon it would be expressed by the equation,

$$CaCO_3 + 2HCl = CaCl_2 + H_2O + CO_2.$$

Careful analysis has shown that marble is almost pure carbonate of calcium. We find, then, in marble two very combustible substances, calcium and carbon, combined with oxygen. Limestone has the same composition. Calcium, then, which is a very great rarity in a free state, is a very abundant element in nature. Its great rarity in a free state is due to its very great combustibility. It is almost impossible to separate it from oxygen, and, when once separated, to keep it from combining with oxygen again.

It has been found that all the rocks and solid matter of the earth are chemical compounds. The rocks are made up chiefly of such combustible elements as potassium, calcium, magnesium, aluminium, and carbon, combined with oxygen. Many of the rarest and most costly metals, then, are the most abundant in nature. The fact that they are so rare in a free state is due to their extreme combustibility.

138. *The Present Materials of the Earth are Products of Combustion.* — We see, then, that water, the most abundant of liquids, is made up of the very combustible elements hydrogen and oxygen; while the solid materials of the earth are composed chiefly of the very combustible elements potassium, magnesium, calcium, aluminium, carbon, and silicon, in combination with oxygen. These materials are the products of combustion. There must have been a time when these elements existed together in a free state. Then, by some means unknown to us, the mass took fire, and the conflagration raged until all the materials were consumed. There was more oxygen than was needed for the combustion, and this is now found in the air in a free state. Oxygen was the most abundant of all the elements, since it alone makes up half the weight of the solid earth, eight ninths the weight of the water, and one fifth the weight of the atmosphere.

Most of the materials of the solid earth are, as has been stated, binary or ternary compounds of oxygen. Some of the metals are found combined with chlorine and sulphur. This is as we should expect, since many of the metals can burn in chlorine or in the vapor of sulphur (136).

139. *Free Oxygen is not necessary to the Support of Combustion.* — We have seen (46) that potassium and sodium will burn on water. Here the combustion is supported by the combined oxygen of the water.

If a mixture of pulverized charcoal and saltpetre be heated in a small crucible, the charcoal burns with great brilliancy. About half the weight of saltpetre is oxygen. When heated with carbon or other combustible substances, it gives up a part of its oxygen very readily. The oxygen thus set free rushes into combination with the combustible substance, causing it to burn vigorously.

Gunpowder is an intimate mixture of pulverized charcoal, sulphur, and saltpetre. A moderate heat causes the carbon and oxygen to combine and the powder to explode.

If zinc be melted in an iron ladle and powdered saltpetre be thrown upon it, the zinc will burn with great brilliancy. Here, also, the oxygen is supplied by the saltpetre.

140. *Reduction of Metallic Ores by means of Carbon.* — The affinity of carbon for oxygen at a high temperature is so great, that this element will burn when strongly heated with the metallic oxides. The carbon in burning combines with the oxygen of the oxide, and leaves the metal in a free state. Advantage is taken of this fact in the *reduction* of metallic ores.

The metals, as has been stated, are found in nature chiefly in combination with oxygen, sulphur, and chlorine. These compounds are called *ores*. The oxides are commonly reduced to the metallic state by heating them with carbon. The sulphides are first roasted; that is, heated while exposed to the air. The oxygen of the air combines with the sulphur of the ore to form a gas called *sulphurous acid*, and the metal also combines with the oxygen of the air to form the oxide of the metal. This oxide is then reduced to the metallic state by heating it with carbon.

The roasting is done in what is called a *reverberatory*

Fig. 39.

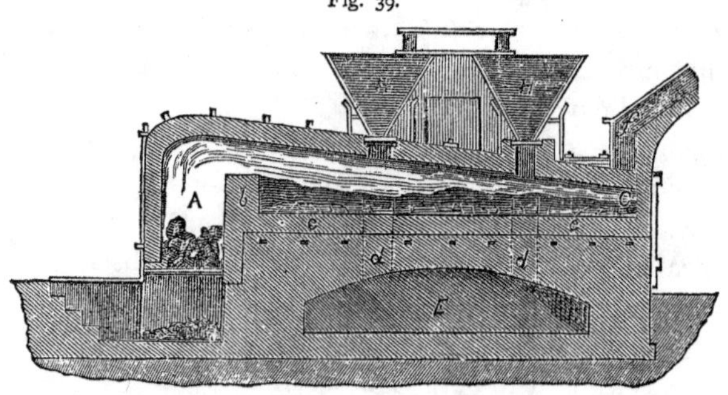

furnace, represented in Figure 39. The ore is put into the hoppers *H H*, from which it falls into the chamber *C*, where it is spread out on the *bed c c*. The fuel is burned on a hearth at *A*, separated from the ore by the bridge *b*. The heated gases rising from the burning fuel are *reverberated*, or reflected, by the arched roof of the furnace, and driven down upon the ore, and then pass off through the flue *f*. When the ore is sufficiently roasted, it is allowed to fall through openings, *d d*, into the chamber *E*. The ore is stirred from time to time to expose fresh surfaces to the action of the air and the flame.

The reduction of the metallic ores is best illustrated by the reduction of iron ore. The *blast furnace* employed for this purpose is represented in Figure 40. These furnaces are usually about fifty feet high, and about fifteen feet in diameter in the widest part of the cavity *C D*. The lowest part, *F*, is called the *crucible*, or *hearth*. *I I* are the *tuyères*, or pipes through which air is forced by powerful bellows. *K K* and *M* are arched galleries for the convenience of workmen employed about the furnace. When working regularly, the furnace is charged from a door at the end of the gallery near the top, first with coal, and then with a mixture of roasted ore and limestone broken into

Fig. 40.

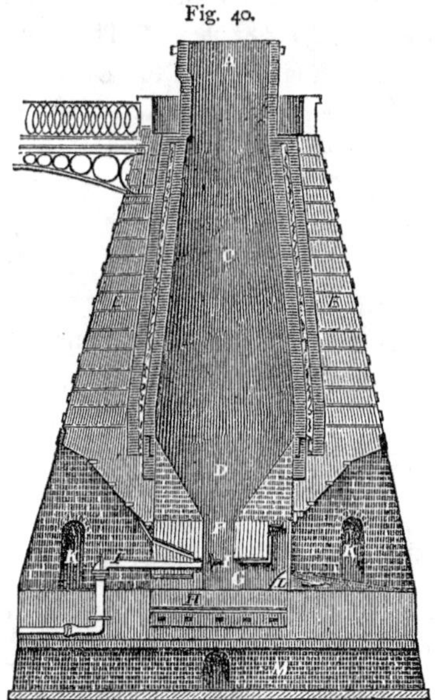

small pieces. As the fuel burns away and the materials gradually sink, fresh supplies of fuel and of ore are added; so that the furnace is kept filled with alternate layers of each.

The oxygen of the air from the bellows combines with the carbon of the fuel, forming carbonic oxide, which rises through the porous mass, and, taking the oxygen from the ore, becomes converted into carbonic acid. The iron mixed with the earthy matter of the ore settles down into the hottest part of the furnace, where both are melted. The iron, being the heavier, sinks to the bottom, where it is drawn off at intervals through a *tap-hole* in the floor *H.* The lighter earthy matter, or *slag*, floats on the surface of the iron, like oil on water, and flows off through an opening above the *tymp-stone L.* The limestone aids in liquefying the earthy matter, and unites with it to form the slag.

The separation of metals from the ores, then, is seen to be a process of combustion. If the ore be a sulphide (or an arsenide) of the metal, this is first burned at the expense of the oxygen of the air, and the sulphur (or arsenic) is converted into sulphurous (or arsenious) acid, and the metal into an oxide.

This oxide (or the original ore, if it is an oxide) is mixed with carbon and heated, and the carbon burns at the expense of the oxygen of the ore, and the metal is left free.

141. *Slow Combustion.* — We have already seen that when iron and other metals are burned in the air or in oxygen, they are converted into oxides. If potassium or sodium be exposed to the air, they soon become coated with a white solid, which is found to be the same as is formed when they are burned in oxygen.

This change, or *oxidation*, has taken place quietly, without development of light, and apparently without development of heat.

So, too, when iron is exposed to moist air, it becomes

covered with a grayish red film, which resembles the powder which collects on the sides of the jar when iron is burned in oxygen. It is, in fact, the same substance. Here, too, the oxidation takes place quietly, and no light is developed, and apparently no heat.

This very gradual burning of a substance without development of light, is called *slow combustion.* When the metals are thus slowly *oxidized*, the process is called *rusting*, and the oxide formed is called *rust.* All the familiar metals, except silver, gold, and platinum, are tarnished on exposure to the air; that is, they become covered with a film of rust, or oxide.

142. *Rusting is attended with Development of Heat.*—This slow oxidation of the metals, as has been stated, takes place apparently without development of heat. But heat is really developed during rusting, though usually so slowly that it is not perceived. If a large pile of iron-filings be moistened and exposed to the action of the air, they rust rapidly, and the temperature rises perceptibly.

A remarkable case of heat developed by rusting occurred in England during the manufacture of a submarine electric cable. The copper wire of the cable was covered with gutta-percha, tar, and hemp, and the whole enclosed in a casing of iron wire. The cable as it was finished was coiled in tanks filled with water; these tanks leaked, and the water was therefore drawn off, leaving about 163 nautical miles of cable coiled in a mass 30 feet in diameter (with a space in the centre 6 feet in diameter) and 8 feet high. It rusted so rapidly that the temperature in the centre of the coil rose in four days from 66° to 79°, though the temperature of the air did not rise above 66° during the period, and was as low as 59° part of the time. The mass would have become even hotter, had it not been cooled by pouring on water.

143. *Decay is Slow Combustion.* — Light a splint of

wood and hold it while burning under a funnel held in an inverted bottle. After a little time pour lime-water into the bottle and shake it. The lime-water becomes milky-white; showing that carbonic acid has been produced by the burning of the wood.

Hold a cold glass vessel over burning wood, and moisture collects on the inside; showing that water is produced by the combustion.

Carbonic acid and water, then, are two of the products of ordinary combustion. Nitrogen has been found to be another product.

Put some peas in a flask, cover them with water, and connect the flask by a glass tube with a second containing a little water. The end of the tube dips under the water in the second flask. Let the flasks stand for some time in a moderately warm place, and bubbles of gas will be seen to escape into the second flask. If this gas is tested with lime-water, it is found to be carbonic acid.

Fig. 41.

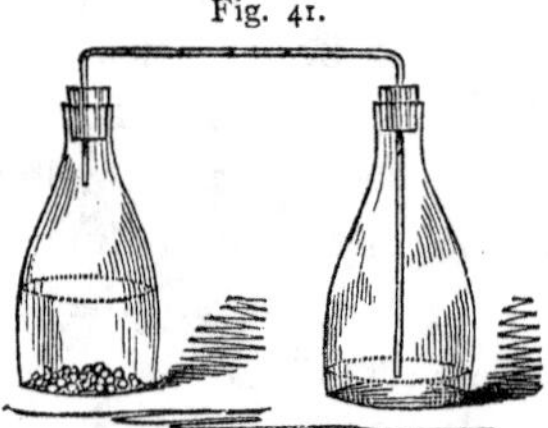

After the peas have soaked a considerable time in the water, remove them and add hydrochloric acid to remove all vegetable matter. Filter a portion of the liquid into a test-tube, add oxide of sodium, and cover the mouth of the test-tube with moistened red litmus-paper. Heat the liquid, and the litmus-paper becomes blue; showing that the liquid contains ammonia.

Carbonic acid and ammonia, then, are two of the products of the decay of the peas. Water is known to be another product.

We see, then, that the products of the ordinary combustion and of the decay of vegetable substances are nearly identical; — the former being chiefly carbonic acid, water, and nitrogen; the latter, carbonic acid, water, and ammonia. Hence we conclude that combustion and decay are analogous processes; decay being, in fact, a kind of slow combustion.

144. *The Decay of Vegetable Substances develops Heat.* — The oxidation of dead vegetable substances takes place so slowly at the ordinary temperatures, that the heat developed is not perceptible. When, however, charcoal which has been finely pulverized for making gunpowder is exposed in large heaps, the oxygen of the air combines with it slowly at first; but, as the heat developed accumulates, the oxidation becomes more rapid, until in some cases the mass takes fire and burns.

So too, when cotton or tow, which has been used for wiping machinery, and has become saturated with oil, is laid aside in heaps, it begins to oxidize slowly; but the heat developed causes the combustion to go on more and more rapidly, until sometimes the heap bursts into a flame.

This rapid combustion, developed gradually from slow combustion, is called *spontaneous combustion.*

RESPIRATION.

145. We have seen that, at a high temperature, vegetable substances combine rapidly with the oxygen of the air, developing intense light and heat, and giving rise to the gaseous products water, carbonic acid, and nitrogen; and that dead vegetable substances, at the ordinary temperatures, combine slowly with the atmospheric oxygen, developing no light and little heat, and giving rise to the gaseous products water, carbonic acid, and ammonia.

But it is well known that a large amount of vegetable matter is consumed as food by man and other animals. What becomes of this vegetable matter?

146. *Animals in Breathing remove Oxygen from the Air.* — It has been found that if a small animal, as a rat or a mouse, is placed on a float on the surface of water, and a bell-jar of air is placed over it, the water slowly rises into the jar; showing that the animal in breathing removes something from the air. After a time, the animal will die; showing that the air has become unfit for respiration. It has also been found, that, if a burning candle be put under the bell-jar with the rat, he will not live half so long as when under the jar alone; showing that animals in breathing remove the same substance from the air as bodies in burning, namely, *oxygen.*

147. *Carbonic Acid is a Product, but not the only Product, of Respiration.* — If we breathe through a glass tube into lime-water, the liquid becomes milky. Hence, carbonic acid is one of the products of respiration.

It has been seen that carbon, when burnt in oxygen, gives rise to carbonic acid. Now it has been found that, when carbon is burned in a jar of oxygen, the volume of the gas is neither increased nor diminished; that is, when oxygen and carbon are converted into carbonic acid, the carbonic acid occupies just the same space as the oxygen which it contains occupied when in a free state. But when a rat breathes in confined air, it has been seen that the volume of the air is diminished. It is evident, then, that the oxygen removed from the air in respiration does not all reappear in the carbonic acid sent out in the breath.

A part of the oxygen taken from the air appears in *water*, which is a second product of respiration; as may be shown by breathing upon any cold substance.

A portion of the products of respiration are removed from the body in a liquid state, by the kidneys. The liquid is found to be highly charged with *urea*, a compound which contains nitrogen, and which, by the action of the air, gives rise to ammonia and carbonic acid.

148. *The Products of Respiration are nearly the same as those of the Combustion and the Decay of Vegetable Substances.* — We see, then, that of the three chief products of the rapid combustion of vegetable substances, of the decay of such substances, and of respiration, two are identical, namely, *water* and *carbonic acid;* while the third product, in the first of these processes, is *nitrogen* ; in the second, *ammonia*, a compound of nitrogen and hydrogen ; and in the third, *urea*, a compound of carbon, oxygen, hydrogen, and nitrogen.

149. *Respiration is a Slow Combustion.* — Respiration is thus seen to be a process of slow combustion, and the warmth of the body is due to the heat developed by this combustion.

COMPOSITION OF VEGETABLE AND ANIMAL SUBSTANCES.

150. From the products of the different kinds of combustion of vegetable substances, it is clear that the three elements, carbon, hydrogen, and nitrogen, are found in these substances.

The presence of hydrogen and nitrogen may be shown by heating some vegetable substance, as beans or macaroni, in a test-tube, with hydrate of sodium or hydrate of potassium. If moistened red litmus-paper is held at the mouth of the test-tube, it is at once turned blue. This proves that ammonia has been produced from the vegetable matter. Ammonia, as we know, contains hydrogen and nitrogen; hence the vegetable matter must have contained those elements.

The existence of carbon and oxygen in vegetable substances may be shown by taking any one of the three most abundant vegetable substances, — starch, sugar, and woody fibre, — and heating it in a test-tube, closed by a cork, with a glass tube passing through it, and conducting the gas produced into a bottle of lime-water. The latter becomes milky, showing that carbonic acid has come from the test-tube.

As the same product is obtained when the test-tube is first filled with nitrogen instead of air, we must conclude that both the carbon and the oxygen of the carbonic acid came from the vegetable matter.

Hence, vegetable substances contain four elements, carbon, hydrogen, oxygen, and nitrogen. They are composed chiefly of the three, carbon, hydrogen, and oxygen. Woody fibre, starch, and sugar contain only these three elements ; but we find in every part of plants small quantities of another class of compounds, containing the four elements just mentioned, and called *nitrogenous* or *nitrogenized* compounds, from the fact that they contain nitrogen. They are also sometimes called *quaternary* compounds (that is, compounds made up of *four* elements), to distinguish them from the other, which are *ternary* compounds.

The nitrogenized compounds are found most abundantly in the *seeds* of plants, such as peas, beans, grain, etc.

It appears, then, that in *rapid combustion* the destruction of vegetable substances is most complete. All the carbon and the hydrogen are separated from the nitrogen, and converted into carbonic acid and water.

In the case of *decay*, the destruction is less complete. All the carbon is separated from the nitrogen and converted into carbonic acid, and all the oxygen also is separated from the nitrogen; but a part of the hydrogen remains combined with the nitrogen as ammonia.

In *respiration*, the destruction is even less complete, since portions of the carbon, hydrogen, and oxygen remain combined with nitrogen in the form of urea.

151. *Animal Substances also contain Carbon, Hydrogen, Nitrogen, and Oxygen.* — Animal substances are also found to contain the same four elements which exist in vegetables.

This is what we should expect, since all animals live either upon vegetables, or upon the flesh of other animals which have lived upon vegetables.

Animal substances differ from vegetable, in being mostly *quaternary* compounds. Nitrogen is one of the most abundant elements in animal compounds.

CONDITION OF OXYGEN DURING DECAY AND RESPIRATION.

152. We have seen that the oxygen of the air at ordinary temperatures is wholly passive, and that during rapid combustion it is wholly active, and that it is roused to activity by means of heat. During decay and respiration the oxygen of the air is in an intermediate state, or one of *partial* activity.

153. *Moisture is necessary to the Beginning of Decay.* — How then is oxygen roused to this state of partial activity? It is found that, when wood is kept perfectly dry, it may be preserved for any length of time without showing the least disposition to decay. Moisture is necessary to cause the decay to begin. The nitrogenized compounds, which exist in small quantities in all parts of vegetables (150), are so unstable, that the mere presence of moisture is sufficient to cause them to break up into simpler compounds. The oxygen of the air takes no part in this process, but the molecules of oxygen, in contact with the compounds, appear to be rendered partially active by their breaking up. Being thus made active, they combine with the carbon and hydrogen of the *ternary* vegetable compounds, and the decay begins. Once begun it is self-sustaining, like ordinary combustion (131). The molecules of the oxygen about the decaying substance are rendered partially active by the process of decay. So, too, in respiration, the oxygen seems to be roused to partial activity by the slow combustion taking place within the blood.

154. *Ozone.* — If one or two sticks of phosphorus, carefully scraped clean, are put into a jar (which must be perfectly free from grease) and partially covered with water, they are slowly oxidized. If the jar be loosely covered and allowed to stand for two or three hours, and the air inside be then examined, it will be found to have undergone a remarkable change.*

Dip a piece of unsized paper first into a thin solution of starch, and then into a solution of iodide of potassium. If this paper be exposed to the air for hours, its color will scarcely change at all; but if it be moistened and introduced into the jar in which the phosphorus has been undergoing slow combustion, it instantly becomes of a deep blue color.

If a piece of silver-foil be put into a jar of this air, it soon crumbles with rust, though silver is scarcely tarnished by long exposure to ordinary atmospheric oxygen.

Hence, we see that the oxygen inside the jar has been rendered partially active by the slow combustion of the phosphorus. Oxygen in this state is called *ozone.*

There are several ways by which oxygen can be converted into ozone, but the most common way is by slow combustion. It would seem that during all processes of decay oxygen is converted into ozone. In what way ozone differs from ordinary oxygen has not yet been well established.

155. *Allotropic States.* — There are several elements which can exist in states wholly unlike in appearance and in many of their properties.

No three substances could be more unlike in appearance than diamond, graphite, and charcoal. Yet each of the three is nothing but carbon, as can be proved by burning each in oxygen gas,† when they all give rise to one and the same product, carbonic acid.

So, too, when sulphur is heated, it first melts into a lim-

* See Appendix, 26. † See Appendix, 27.

pid, light-colored liquid; then becomes dark and thick; and then light and limpid again. If now it be suddenly cooled, by pouring it in a fine stream into cold water, it becomes soft and plastic, like wax, while ordinary sulphur is hard and brittle. If left to itself, this plastic sulphur returns after a time to its ordinary hard and brittle state.

These different states of the same element are called *allotropic* states.

The fact that the same element can exist in states so wholly unlike, is attracting a good deal of attention among scientific men. It seems to point to the conclusion that all the elements may after all be only allotropic states of a single element, or at most of a few elements.

THE GROWTH OF PLANTS.

156. *The Constituents of the Air are Constant.* — Notwithstanding that enormous quantities of oxygen are daily removed from the air in the processes of combustion, decay, and respiration, and that the oxygen is in each of these cases replaced by carbonic acid and watery vapor, the relative quantities of oxygen, carbonic acid, and watery vapor in the atmosphere are nearly constant. The amount of each of these substances found in the air in different places, and at different times, varies but slightly. How is this to be explained?

157. *Rain and Dew.* — The amount of watery vapor which the air at any given temperature can hold is limited. When the air has taken up all the vapor it can hold, it is said to be *saturated.* As soon as the air has become saturated, the watery vapor begins to return to the liquid state in the form of rain and dew. It is thus prevented from accumulating to any great extent in the atmosphere.

158. *Plants in growing remove Carbonic Acid from the Air, and replace it with Oxygen.* — If a leafy plant be placed un-

der a glass vessel and set in the sunshine, and a stream of carbonic acid be made to pass slowly over it, and the gas, as it escapes from the vessel, be collected and examined, it will be found that a part of the carbonic acid has been removed and replaced by oxygen.

While, then, the destruction of plants, by the various processes of combustion, decay, and respiration, is removing oxygen from the air and pouring carbonic acid into it, plants in growing are silently removing this carbonic acid from the air and replacing it with an equal bulk of oxygen; and since plants grow as rapidly as they are destroyed, the relative quantities of oxygen and carbonic acid in the air remain unchanged.

159. *The Growth of Plants is a Chemical Process.* — We have seen that growing plants are continually removing carbonic acid from the air, and giving out oxygen. We conclude, then, that plants derive their carbon from this carbonic acid. In the leaves of the plant, under the influence of sunlight, the carbonic acid is decomposed, the carbon stored away in the plant, and the oxygen given back to the air.

We have already noticed that the most abundant metals are very rare and costly substances in their free state, owing to the difficulty of separating them from the oxygen or chlorine with which they are combined. And it is only by the aid of such rare metals as potassium and sodium, which have a very strong affinity for oxygen, that we can separate carbon from its oxygen in carbonic acid. Hence, if we were obliged to obtain carbon by separating it from its combination with oxygen, this useful element would be as rare and as costly as potassium and sodium. And at best we can obtain carbon from carbonic acid only by removing the oxygen and locking it up in combination with some element which has a stronger affinity for it. The chemist knows no way of separating the carbon, and letting the oxygen go free. But this is continually done by the sunbeam in the delicate laboratory of the leaf.

Plants thus obtain their carbon from the carbonic acid of the air. It is pretty well established, also, that they obtain their hydrogen, oxygen, and nitrogen from the two compounds water and ammonia, which, as we have seen, are products of their decay. The ammonia is washed out of the air by the rain, and, thus dissolved in water, is taken up by the roots of plants, and conveyed thence to the leaves. It is probable that plants obtain a part of their carbonic acid also through the roots, since this substance is soluble in water.

The growth of plants is thus seen to be a chemical process, just the opposite of that of their decay. The three compounds carbonic acid, water, and ammonia are decomposed in the leaves of the plant; a part of the oxygen is given back to the air; and the remainder of the oxygen and the other elements are rearranged so as to form the various vegetable compounds which serve as food for man and other animals, and as fuel for our fires.

160. *Nitrogen in the Air.* — The oxygen may be removed from the air by means of phosphorus, as we have seen, or by passing the air through a red-hot iron tube filled with copper turnings. At this high temperature the copper combines with the oxygen. If the air from which the oxygen has been removed be passed through lime-water, and then through a tube filled with fused chloride of calcium (59), the carbonic acid and the watery vapor will be taken from it. After doing this we still have about four fifths of the air left, and this large remainder we find to be pure nitrogen. We have already seen (129) that about one fifth of the air is oxygen. It is evident, then, that there is but a small amount of watery vapor and carbonic acid in the air at any one time, though immense quantities of both are continually passing through it.

161. *Why Nitrogen can exist in the Air in a Free State.*— It is at first surprising to find a free element in the air along with oxygen, especially when we remember that the present materials of the earth are the result of a great conflagration, in which oxygen was the chief agent. But we remember that the leading characteristic of nitrogen is its inertness. It will neither burn itself, nor allow a body to burn in it. It is owing to this inertness that nitrogen can exist in the air in a free state. It is with the greatest difficulty that nitrogen can be made to combine directly with any element except boron. Its inertness also explains the fact that nitrogen is found to so small an extent in the compounds of the earth. It exists in the solid crust of the earth only in the nitrates of potassium and sodium, and these nitrates we find only in small quantities. Nitrate of sodium is found chiefly in Peru, and nitrate of potassium in India. These salts are of the greatest importance, since it is from them that nitric acid and the various nitrates are prepared.*

The chief use of nitrogen is to moderate combustion. Were not the activity of oxygen tempered by the inactivity of nitrogen, it would be impossible to confine or control our fires. Our stoves would burn as readily as our coal now does.

162. *Atmospheric Elements.*—We find in the air four elements, hydrogen, oxygen, nitrogen, and carbon; the oxygen and the nitrogen in a free state, and the hydrogen and carbon in combination. These elements are, therefore, called *atmospheric elements.* It will be noticed that they stand at the head of the four groups of elements (79).

Since these elements mainly make up all animal and vegetable substances (or all *organic* substances, as they are called), they are also known as *organic* elements.

* See Appendix, 29.

SUMMARY.

Ordinary combustion, decay, and respiration are chemical processes, differing from one another mainly in the rapidity and the completeness with which they take place.

The division of substances into *combustibles* and *supporters of combustion* is convenient, though merely conventional, since combustion consists in the chemical union of two or more elements, and in reality one of the elements is just as combustible as another. (128.)

The present materials of the earth, being mainly compounds of hydrogen and carbon with oxygen, and of the metals with oxygen, chlorine, and sulphur, must be regarded as products of combustion. (138.)

The metallic oxides may be *reduced* by heating them in contact with carbon, which burns at the expense of the oxygen of the oxide, forming carbonic acid, and setting the metal free. (140.)

The metals thus set free return to the state of oxides again by *rusting*, which is a process of slow combustion. (141.)

The products of combustion, decay, and respiration are chiefly *carbonic acid*, *water*, and *ammonia*. (148.)

The growth of plants is a chemical process, just the opposite of that of their decay. In growing, plants remove from the air carbonic acid, water, and ammonia. These compounds are broken up; the oxygen in part given back to the air; and the other elements, with the remaining oxygen, rearranged so as to form the various vegetable compounds. (158, 159.)

Animals derive all their food, directly or indirectly, from plants; while plants derive their food, directly or indirectly, from the air. (151, 159.)

Animal and vegetable (or *organic*) compounds contain the four elements carbon, hydrogen, nitrogen, and oxygen. Animal compounds differ from vegetable in being nearly all *quaternary*, while vegetable compounds are chiefly *ternary*. (150, 151).

The atmosphere contains oxygen, nitrogen, carbonic acid, and water. The four elements oxygen, nitrogen, carbon, and hydrogen are consequently called *atmospheric elements*. Since they mainly make up animal and vegetable substances, they are sometimes called *organic* elements. They head the four groups of elements. (162.)

Oxygen exists in three conditions: (1.) wholly passive, (2.) partially active, and (3.) wholly active. Its partially active state is regarded as an *allotropic* state, and is called *ozone*. (152, 154.)

Several elements are characterized by allotropic states. Those of carbon and sulphur are well marked. (155.)

Hydrogen, at a high temperature, combines with oxygen with great energy, as is shown by the violent explosion of a mixture of the two gases, and by the intense heat of the oxy-hydrogen flame. (130.)

Nitrogen is remarkable for its inertness, to which it owes its existence in the atmosphere. It is rarely found in combination in nature. In this respect, it is in striking contrast to oxygen, which is found in combination with most of the elements. (160, 161.)

DESTRUCTIVE DISTILLATION AND ITS PRODUCTS.

163. We have already seen how vegetable substances can be destroyed by the various kinds of combustion.

When vegetable substances are heated in closed vessels, so as to exclude the oxygen of the air, they are broken up into a number of compounds, which vary with the temperature to which they have been exposed. In combustion these organic compounds are broken up by the affinity of oxygen brought to bear upon them from without; while in the other case they are broken up by the internal action of heat. In the first case, new compounds are formed by the addition of new material; in the second case, by subdivision without any addition of new material. The first process is called combustion; the second, *destructive distillation.*

164. *The Preparation of Charcoal.* — One of the simplest cases of destructive distillation is to be seen in the preparation of charcoal. This process may be illustrated by putting pieces of dry wood into a test-tube, closed with a cork, through which a glass tube passes. Heat the test-tube, and the wood turns black. As has already been shown, carbonic acid will pass off (150). Hold a lighted taper to the tube, and the escaping gas takes fire. Hold a cold glass rod in the flame, and it becomes covered with soot, showing the presence of carbon in the flame. Hold a cold glass vessel over the flame, and moisture collects upon it, showing that water is a product of the burning of this gas. Besides the carbonic acid, then, an inflammable gas issued from the tube; and this gas must be a compound of hydrogen and carbon, since carbon is found in the flame, and water is one of the products of the burning gas.

A part of the carbon, then, combines with oxygen to form carbonic acid; another part with hydrogen, to form the inflammable gas; while the greater part remains behind as a black solid. This black residue is charcoal, a form of carbon. If the heating is continued, and the gaseous products are conducted into a cold receiver, a liquid product is also obtained.

One way of preparing charcoal is to place billets of wood in an iron cylinder, which is closed air-tight and heated to dull redness. The volatile products are driven off and allowed to escape through a flue, and the solid charcoal remains behind.

A ruder method is practised in the country, where wood is plenty. A stake is set in level ground, and brushwood heaped around its base. Wood is then stacked round the stake, so as to form a mound some 20 or 30 feet in diameter. This mound is then covered, first with leaves or turf, and then with earth, leaving only a small opening at the bottom, through which the mound is set on fire. When the fire is well under way, the mound is covered more deeply and allowed to burn slowly out. This requires about a month. The burning of a part of the wood furnishes heat for charring the rest.

165. *The Products of the Distillation of Wood.* — When hard wood, as beech, is subjected to destructive distillation in a retort, and the volatile products are condensed in a suitable vessel, four principal classes of substances are formed: (1.) gases; (2.) a watery fluid; (3.) a dark resinous fluid; (4.) charcoal.

(1.) This product is a mixture of inflammable gases, the most important of which are the two *hydrocarbons* (or compounds of hydrogen and carbon) *marsh-gas*, $H_4€$, and *olefiant gas*, $H_2€$.

(2.) This product is an acrid liquid, known as *pyroligneous acid*, or *wood vinegar*. From this acetic acid is obtained, which is used in large quantities in the preparation of the acetates of iron, lead, and soda, which are much employed in dyeing and calico-printing.

(3.) This product is *wood tar*, a thick liquid, insoluble in water, but soluble in alcohol. Its chief use formerly was for tarring and calking ships, but recently it has become an important source of both illuminating and lubricating oils. These oils will be more fully treated hereafter.

(4.) This product is the charcoal remaining in the retort. It is used chiefly as fuel and in reducing metallic ores.

166. *Ingredients of Wood Tar.* — When beech-wood tar is distilled, a light oil passes over first, called *eupion*, or *wood naphtha.* It is now often sold under the name of *benzole*, and used as a burning-fluid, for removing oil-stains from clothes, and for countless other purposes. It burns with a brilliant white flame, free from smoke; but its extreme inflammability makes it a dangerous liquid for lamps.

After this light oil has distilled over, a heavy oil follows. It contains various ingredients, the chief of which are *creosote* and *paraffine.*

167. *Creosote.* — This is an oily, colorless liquid, with a peculiar smoky odor. It has remarkable *antiseptic* (or preservative) properties. A piece of flesh steeped in a very dilute solution of it dries up into a mummy-like substance, which refuses to decay. Meat, as tongues or hams, may be almost instantly *cured* by dipping it into a solution containing one part of creosote to 100 parts of water or brine. It is this substance which imparts to wood-smoke its property of preserving meat. It is a compound of carbon, hydrogen, and oxygen.

168. *Paraffine.* — This is a pearly-white, tasteless, and odorless solid. The most corrosive acids and alkalies have no effect upon it. Hence its name, from *parum*, little, and *affinis*, from which *affinity* is derived.

It burns with a bright white flame, without smoke. It is

now much employed as a material for candles, which for purity and lustre are not surpassed by even the best and most costly wax-candles.

Unsized paper, after having been soaked in paraffine, may be kept for weeks in concentrated sulphuric acid without undergoing the slightest alteration. Hence it is an excellent coating for the labels of bottles in which acids are kept.

169. *Asphalt.* — Asphalt, or pitch, is the residue left after distilling tar. It is used for varnishes, and as a material for making lamp-black.

170. *Products of the Decay of Vegetable Substances when Air is excluded.* — When vegetable substances are consumed by the slower process of decay, with a partial or complete exclusion of air, the products are somewhat different from those of ordinary decay. The gas obtained by stirring the mud in marshes and at the bottom of stagnant pools (72) is formed in this way, and is made up chiefly of marsh gas, H_4C, and carbonic acid, CO_2.

171. *The Formation of Mineral Coal.*—In tropical swamps where vegetation is rank, vast masses of vegetable matter accumulate, and gradually decay under water. In some cases the land at the bottom of these swamps is slowly sinking, and the bed of peat, as it sinks with it, becomes covered with mud and sand, which numerous streams are washing down upon it. This goes on, year after year and century after century, until the bed is buried hundreds of feet beneath the surface. The vegetable matter thus sunk in the earth, and subjected to enormous pressure, gradually undergoes a process of internal combustion similar to that which takes place under water. In many cases the decomposition is hastened by the agency of the internal heat of the earth. It is probable that the vast beds of coal found in various parts of the earth have been thus formed. All this coal is the remains of an ancient vegetation, and it

undoubtedly required millions upon millions of years to complete its conversion into coal.

172. *Hard and Soft Coals.* — The mineral coals may be conveniently divided into hard, or *anthracite*, and soft, or *bituminous* coal, and there are several varieties of each.

The main differences between the two are these: hard coal is almost pure carbon, while soft coal contains also considerable hydrogen and some oxygen; hard coal still retains the cellular structure of the wood, which is clearly seen under the microscope, while in soft coal this cellular structure is almost entirely wanting; hard coal burns without flame, soft coal with flame.

It is found by experiment that, if vegetable matter be enclosed in an apparatus made of wet clay, and subjected for a long time to great pressure and to a high temperature, a variety of coal will be formed closely resembling hard coal. In this case the gases which are formed escape through the porous clay. If strong iron cylinders are used instead of clay, a variety of coal resembling soft coal is formed. In this case the gases have no means of escape. In the first case the coal retains the cellular structure of the wood; while in the second case this cellular structure is entirely destroyed, and the carbon appears to have been dissolved in the liquid and gaseous products formed at the same time.

It is probable that, in the slow decomposition of the vegetable matter buried in the earth and subjected to great pressure, both these conditions have existed. In some cases the gaseous products were free to escape as they were formed, and hard coal is the result; in other cases the gases could not escape, and bituminous coal, or soft coal, is the result. This bituminous coal seems once to have been in a liquid state, and afterwards to have hardened. If, when in the liquid state, it was surrounded by porous strata, it was absorbed by these and afterwards hardened. This seems to have been the origin of the bituminous shales and slate which occur in immense beds in various parts of the earth. These are loose, clayey rocks, impregnated with bituminous matter.

173. *Products of the Distillation of Bituminous Coal.* — The products obtained by the destructive distillation of coal are still more numerous than those obtained from wood. Wood, containing much oxygen and comparatively little nitrogen, furnishes compounds which contain much acetic acid and little ammonia, and which, therefore, have an *acid* reaction. Coal, on the other hand, contains much nitrogen and little oxygen, and gives products rich in ammonia, and having consequently an *alkaline* reaction.

When coal is distilled at high temperatures, an abundance of an inflammable gas is obtained, and also a large amount of liquid products, which are then called *tars*.

When coal is distilled at a low temperature, but little gas is obtained, and the liquid products are then called *oils*.

174. *The Composition of Coal-Tar.* — Coal-tar has been found to contain three classes of substances: —

(1.) Acid oils, soluble in alkalies.

(2.) Alkaline oils, soluble in acids.

(3.) Neutral oils, not affected by alkalies and some acids.

(1.) The most important and abundant ingredient of the acid oils is carbolic acid, C_6H_6O. It is analogous to creosote, and is sometimes called *coal creosote*. It is now largely employed as a permanent dye-stuff for silk and woollen goods. This dye-stuff is prepared by heating carbolic acid moderately with nitric acid. This mixture is called picric acid. On evaporating this liquid, yellow scaly crystals are obtained. Like all the tar colors, its dyeing qualities, when in solution, are most intense. Silk and woollen goods put into the solution, even when cold, assume a rich yellow color, far surpassing that obtained from other dyes.

(2.) The alkaline oils constitute but a small fraction of the tar. Their most important ingredients are ammonia and *aniline*.

Aniline is an oily substance which, when acted upon by compounds which readily part with oxygen, furnishes a complete series of the most brilliant dyes. The preparation of these rich dyes from aniline is one of the most interesting discoveries of modern times, and has caused almost a revolution in the arts of dyeing and calico-printing. It is still more surprising when we consider that these brilliant colors are obtained from what was until recently a disagreeable waste product of the gas-works. When first prepared, they were worth their weight in gold ; now, they can be bought at a comparatively moderate price.

Their dyeing qualities are so intense that a little material goes a great way ; so that, notwithstanding their high price, they are more economical than any ordinary dye-stuffs.

(3.) The neutral oils are the *coal-oils* proper. They contain a great variety of compounds, both liquid and solid, the latter being held in solution. Of the liquids, *benzole*, *toluole*, and *cumole* are the most important ; and of the solids, *paraffine* and *naphthaline.*

175. *Benzole and Nitro-benzole.* — Benzole, which has been already mentioned (166), is a very important compound, as it is the material from which aniline is usually prepared. The symbol for benzole is C_6H_6. When mixed with concentrated nitric acid, it is converted into *nitro-benzole*, $C_6H_5NO_2$.

$$2C_6H_6 + N_2O_5 = 2C_6H_5NO_2 + H_2O.$$

Nitro-benzole is the artificial *oil of bitter almonds*, and is much used in the art of perfumery. Its most important use, however, is in the preparation of aniline. When heated with acetic acid ($C_2H_4O_2$) and iron filings, it loses two atoms of oxygen and takes up two of hydrogen, and is converted into aniline.

$$C_6H_5NO_2 - O_2 + H_2 = C_6H_7N \text{ (aniline)}.$$

176. *Toluole and Cumole.* — Toluole, $€_7H_8$, and cumole, $€_9H_{12}$, are the chief ingredients of the well-known illuminating or lamp oils obtained from coal.

177. *Naphthaline.* — Naphthaline is a beautiful, pearly-white solid. It is inflammable, but burns with a smoky flame and a disagreeable odor. Brilliant red and blue colors, rivalling those prepared from aniline, have lately been obtained from this solid.

When vegetable matter is distilled at a high temperature, benzole and naphthaline are formed in great abundance, with but small quantities of toluole, cumole, and paraffine. When, on the other hand, the distillation is conducted at a low temperature, toluole, cumole, and paraffine are formed in large quantities, with but little benzole and paraffine.

COAL-OILS.

178. At the beginning of the present century, the means of lighting our dwellings consisted, in the main, of poor tallow-candles and dim and dirty oil-lamps. On the continent of Europe, whale-oil and other animal oils were costly, and there consequently resort was had to natural tar and bituminous slate, in order to obtain illuminating oils. For more than twenty years past, lamp-oils have been extensively prepared from wood, rosin, and bituminous matter.

In Great Britain and in this country the manufacture of coal-oils is of much more recent growth, because the extensive whale-fisheries supplied all the wants of the market.

The manufacture of coal-oil was introduced into this country in 1853, and was at first confined to those districts where bituminous coal could be mined at a cheap rate.

Soon after this manufacture was established in this country, and after the value of coal-oils came to be fully recognized, attention was drawn to *petroleum*, or rock-oil, as a ready means of supplying these oils cheaply. On examination this oil was found to be analogous, in its composition and its properties, to that obtained from bituminous materials.

179. *The Origin of Petroleum.* — We have seen (172) that, in the original decomposition of the vegetable matter of former ages, portions of it were probably reduced to a liquid state, and afterwards hardened into bituminous coal. When in this liquid state, it must have closely resembled petroleum. We have also seen that petroleum closely resembles the coal-oils obtained by the destructive distillation of soft coal and other bituminous substances. All scientific men are agreed that the petroleum found in the earth results from the decomposition of organic matter, and nearly all are agreed that it results mainly from the decomposition of *vegetable* matter. It is, however, a disputed point, whether it results from the original decomposition of the vegetable substances, or from the action of the internal heat of the earth on the bituminous coal at a subsequent period.

It is probable that the petroleum now found in the earth is the product both of the original decomposition and of subsequent distillation. Petroleum is, however, rarely found in contact with bituminous strata of any kind. It is more often found in fissures in sand rocks; rocks in which no oil could ever have been generated, since whatever organic matter they might have contained was too much exposed to atmospheric oxygen to admit of its being *bitumenized*, or made bituminous. It is not only impossible that the oil could have originated in these sand rocks, or in the sandy shales which underlie them in the Oil Region in Western Pennsylvania, but it is most probable that the oil ascended from still lower rocks in the form of vapor, which condensed in the cavities above. Since, then, petroleum is seldom found where it originated, but ordinarily in cavities of rocks higher up, it seems probable that it is *mainly* the product of distillation.

The chemical conditions essential to the generation of oil have evidently existed over a very wide area; but the oil is found only where fissures exist in the rocks. These fissures serve two purposes: one, to give space for the formation and expansion of the hydrocarbon vapor; the other, to furnish receptacles for the condensed oils. These fissures must connect with the sources of the oil. If they have any outlets at the surface of the earth, by which the more volatile portions of the oil may escape as gas, the oil within them becomes thicker and heavier. Hence, as a general rule, the oil found near the surface is heavy, the cavities containing it being likely to have outlets. It may, of course, happen that a deeply-seated fissure has such an outlet.

180. *How Petroleum is obtained.* — The oil is obtained by piercing one of these cavities by a well. It often happens that the upper part of the cavity is filled with pent-up uncondensible gases. In this case, if the well happens to pierce the lower part of the cavity, the expansive force of the confined gases will drive the oil from the well in a continuous stream. Oil is often forced from a new well with such velocity, that it rises in a jet a hundred feet high.

It sometimes happens that the lower part of the cavity is filled with brine, upon which the oil floats. If, in this case, the well pierces the lower part of the cavity, brine is the first product. After a time the salt-well may change to an oil-well.

COAL-GAS.

181. *The History of Gas-lighting.* — The idea of turning hydrocarbon gases to the practical purposes of illumination occurred at about the same time to Murdock, in England, and Lebon, in France.

As early as 1691, Dr. Clayton discovered that an inflammable gas could be obtained from the destructive distillation of coal; but no one thought seriously of using the gas for illumination till about 1790, when Mr. William Murdock, afterwards connected with Messrs. Bolton and Watt's engineering workshops at Soho, turned his attention to this subject.

In 1792, Mr. Murdock lighted his own house and office with gas which he distilled from coal. In 1802 he made a public exhibition of gas-lighting at the Soho foundry; and Mr. Matthews, an eyewitness, says: "The illumination of the Soho works on this occasion was of the most extraordinary splendor; the whole of the front of that extensive building was ornamented with a great variety of devices, that admirably displayed many of the various forms of which gas-light is susceptible. This luminous spectacle was as novel as astonishing, and Birmingham poured forth its numerous population to gaze at and admire the wonderful display of the combined effects of science and art." The subject now gained the attention of other eminent scientific and practical men. But for several years only private gas-works were erected, for cotton-mills and similar establishments.

In 1808, Mr. Clegg, an able mechanic, to whose ingenuity are due many of the main features of the present system of gas-lighting, first introduced the method of purifying the gas by passing it through milk of lime.

In 1804, Mr. Winsor endeavored to form an incorporated company for the full development of gas-lighting, so that the streets, shops, and private dwellings should enjoy its advantages. "But gas-lighting, like every other great innovation, was looked upon by the public with excessive distrust. In the event of its success, several branches of industry and commerce were doomed to suffer; many interests were supposed to be at stake; some of the chemical properties of the gas were unknown; great doubts existed as to its safety, and fears as to its salubrity; indeed, the danger of explosion was magnified to the extent that it was asserted, and believed, that a town could be destroyed by the explosion of the main pipes in the street; and interested parties, in order to prevent the establishment of gas-lighting, did not scruple to appeal to the naval glory of the nation, and this shortly after Nelson had achieved his great victories. 'If,' said the opponents of the new light, 'this becomes successful, then our naval supremacy is gone, for at present we obtain our artificial light principally from the whale fisheries; these are the nurseries of our best sailors; so, if we destroy the one, the other must be affected; if the fisheries no longer exist, our navy must degenerate.' At length, after having struggled, during four years, single-handed, and as it were against the opinion of the world,—having by his letters, pamphlets, and lectures proved the advantages of gas,—Mr. Winsor succeeded, in 1807, in obtaining a capital of £20,000, by means of subscribers, preliminary to the formation of a company."

Mr. Winsor and his subscribers could not obtain a charter of incorporation till 1812. Even then the enterprise was looked upon as so visionary, that the act of incorporation was said to be granted in order to make a great experiment of a plan of such extraordinary novelty. In December, 1813, Westminster Bridge was first lighted with gas. From this time gas-lighting made the most rapid progress in England; and now the consumption of gas in London alone amounts to more than seven billions of cubic feet annually. To make this gas, eight hundred thousand tons of coal are required; while the length of the main pipes in the streets of the city is more than two thousand miles.

Paris was first lighted with gas in 1820. There, as in England, strong prejudices had to be overcome.

182. *Manufacture of Coal-Gas.* — The most essential

parts of the apparatus used in the making of coal-gas are represented in Figure 42.

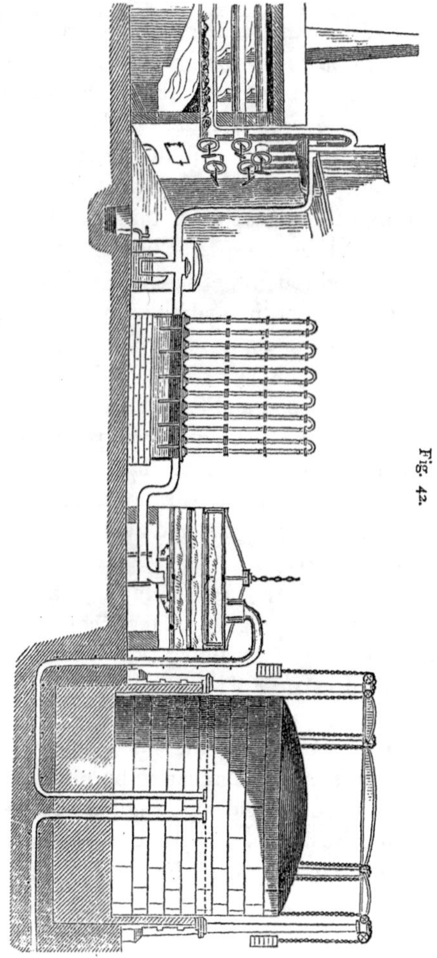

Fig. 42.

Of course, it is only soft or bituminous coal that can be used for making gas. This coal is distilled in long iron retorts, seen at the left of the figure. When charged with coal, these retorts are closed air-tight. They are then heated to a very high temperature by the furnaces beneath.

The gaseous and volatile compounds formed by the distillation of the coal pass up through pipes (one of which may be seen leading from each retort) into a long horizontal pipe, called the *hydraulic main*. This is half full of water; and it will be noticed that the pipes leading from the retorts dip beneath the surface of this water. The gas readily passes from the pipes by bubbling up through the water; but, when it has once passed into the main, it cannot pass back again through the water.

The gas passes on from the hydraulic main, through the pipe leading towards the right, into a tank, called the *tar cistern*. By this time the more condensible gases have returned to the liquid state, and collect in the smaller vessel into which the pipe passes, and from which they overflow into the larger tank. From the latter they are drawn off at intervals. These condensed products are coal-tar and a lighter liquid highly charged with ammonia, and called *ammoniacal liquor*.

The uncondensed gases pass on through the series of upright pipes shown in the figure. Here they become still further cooled, and all the remaining condensible gases are reduced to the liquid state. This system of pipes is called the *condenser*.

After leaving the condenser, the gas still contains, besides the compounds fit for illuminating purposes, the noxious compounds carbonic acid and sulphide of hydrogen, or hydrosulphuric acid (88). These are removed in the next part of the apparatus, called the *purifier*. It consists of a chamber with several perforated shelves, which are covered with slaked lime. In passing over this lime, the carbonic and hydrosulphuric acids are absorbed, while the purified gas passes along into the *gasometer*, or gas-holder, seen at the extreme right of the figure.

This gasometer consists of a large sheet-iron bell-jar, which dips into a cistern of water. The latter is deep enough to allow the bell to be completely submerged, and filled with water. The bell is counterpoised with weights, and rises as the gas passes into it.

From the gasometer the gas passes out, through the pipe shown in the figure, into the streets and houses of the city.

The gas received into each house is made to pass through a self-acting instrument, called a *gas-meter*, by which it is measured. One of the most common forms of the instrument (see Figure 43) consists of an outer case, *a a*, filled with water up to the horizontal dotted line, and, within it, a revolving drum, divided into four compartments, *c*, *d*, *e*, *f*, by as many bent partitions. The bending of the partitions forms a central space, *g*, and thus the gas can pass from one division into the next. The gas enters at the back of the outer case by a pipe which passes into the central space, where it rises a little above the surface of the water. As one compartment gets filled with gas, it becomes lighter and rises, thereby causing the drum to perform a fourth of a revolution. In the figure, the gas is passing into the division *c*. As this fills and rises, *d* comes into the same position; then *e*; and then *f*, which being filled, and rising, completes one revolution. It will be seen that, as each compartment rises above the level of the water, the gas contained in it can pass out through a slit in the rim of the drum into the outer case, and from the top of the case a pipe conveys it to the burners.

Fig. 43.

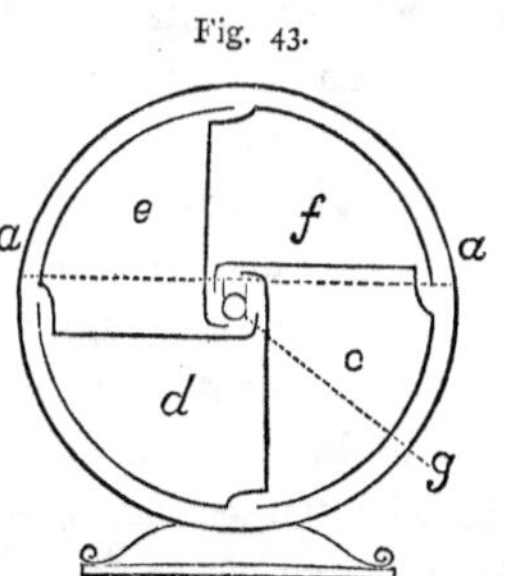

Thus, while one compartment is losing its gas, another one is filling, and so on. The revolution of the drum gives motion to a train of wheels, which in turn move the hands on dial-plates, and thus register the number of cubic feet of gas that have passed through the meter.

ILLUMINATION.

183. *Nature of Flame.* — It has already been noticed that coal-gas burns with flame, while the solid carbon burns without flame. All combustible gases burn with flame.

Many solids, as wood, wax, and tallow, appear to burn with flame.

We have seen that wood, when heated in a closed tube, gives off an inflammable gas. If either wax or tallow be heated in the same way, an inflammable gas is given off. Thus it has been found to be true, that every solid which appears to burn with flame can be converted into an inflammable gas by means of heat. When these solids begin to burn, the heat developed is sufficient to generate this inflammable gas. It is this gas, and not the solid, which burns with flame.

Flame, then, is gas burning.

184. *The Light of the Flame.* — In the oxy-hydrogen flame (130) there is but little light, though an intense heat. To what is the light of the ordinary lamp or gas flame due?

185. *A Solid and Heat are necessary to Illumination.* — We have already seen that the oxy-hydrogen flame develops intense heat, but scarcely any light. If, however, a cylinder of lime is held in this flame, the light becomes most intense. The light obtained in this way is known as the *calcium* light, or the *Drummond* light; and when its rays have been gathered and reflected by a mirror, it has

been seen at a distance of a hundred miles in broad daylight.

We see, then, that a solid and an intense heat are necessary to illumination.

186. *These two Conditions are fulfilled in the Burning of Coal-Gas.* — Illuminating-gas, as we have seen (125), is a compound of carbon and hydrogen. We also found, by means of the cold glass rod, that there is free carbon in the flame. We therefore conclude that the hydrogen of the gas combines with the atmospheric oxygen first. In doing so, it develops great heat.

It has been observed that, in the manufacture of charcoal, the delicate cells of the wood are preserved intact; showing that the carbon, at the high temperature to which it is exposed in the process, has no disposition to melt. Carbon is equally infusible in the highest temperature that we can produce. Its fixedness in the solid state is its most marked characteristic.

When the hydrogen of coal-gas combines with the oxygen of the air, the carbon is set free in a solid state. As the hydrogen consumes all the oxygen in the nearest layer of air, the carbon must delay a moment in the flame before it can get at oxygen to unite with. This finely-divided solid thus becomes heated white-hot, and gives out light. This condition lasts but an instant. The next instant the carbon combines with atmospheric oxygen, and passes off as a colorless gas. As fast, however, as the carbon is consumed, a fresh supply is set free, and thus the light is constantly kept up.

The value, then, of coal-gas for illumination depends upon a most delicate adjustment of affinities.

Were the affinity of carbon for oxygen a little stronger than it now is, the carbon would burn at the same time with the hydrogen, and there would be no light in the flame. On the other hand, were the affinity of carbon for oxygen less than it is, the carbon would not burn at all, but after developing light would pass away from the flame in the form of soot.

187. *Bunsen's Lamp.* — The fact that carbon must remain an instant in the flame in the solid state to develop light, is illustrated by Bunsen's burner. It is shown in Figure 44, and consists of a brass tube, near the bottom of which are four round holes, through which the air can pass into the tube. A second tube opens into the inside of the brass tube from the bottom, just on a level with these holes. The coal-gas passes into the larger tube through this second tube. The air passes in through the holes at the same time, and the two gases become intimately mixed before they reach the top of the tube. Upon lighting the mixture as it escapes from the tube, it burns with scarcely any light, but a good deal of heat. If the holes at the bottom of the tube are closed, the gas burns with the ordinary luminous flame. In the first case, there is an excess of oxygen mixed with the coal-gas, so that there is enough to combine with the carbon as soon as it is set free, before it becomes sufficiently heated to develop light. In the second case, the carbon cannot get at the oxygen to combine with it, until it has passed through the burning layer of hydrogen, and thus become intensely heated.

Fig. 44.

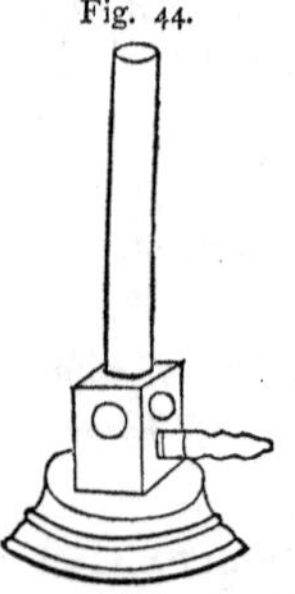

188. *The Best Shape for a Gas-Flame.* — From what has already been said, it is evident that the light of the flame is developed only at the surface, in the burning layer of hydrogen. The ordinary gas-burner is so made that the flame will be flat, in order that it may have as much surface as possible. The greater the surface, the greater the light developed.

189. *The Argand Burner.* — A sheet of paper will evi-

dently present to the air just as much surface when it is bent round till the two edges meet, as when it is flat; while in the second case it is in a more compact form. So the ordinary gas-flame will present just as much surface, if its edges are bent round till they meet. In the Argand burner, this cylindrical form is given to the flame, by supplying the gas through a circle of small holes in a hollow brass ring. A current of air passes up through this ring, furnishing oxygen to the inner surface of the flame.

190. *The Bude Light.* — If the interior of the ring of an Argand burner be closed, the flame becomes of a dull red color. Since no oxygen is supplied to the inner surface of the flame, the combustion is imperfect, and develops but a low degree of heat.

If, on the other hand, a stream of pure oxygen is supplied to the interior of the cylindrical flame, the flame diminishes in size, but the light becomes very intense, and of a pure white color. The increased supply of oxygen increases the energy of the combustion of the hydrogen, and the intensity of the heat developed by it. In the first case the carbon is heated only to a dull redness, while in the second case it is raised to a full white heat.

191. *A Burning Candle is a miniature Gas-Factory.* — We have already seen that wax or tallow, when heated in a flask, is converted into an inflammable gas. This gas in burning produces water and carbonic acid; hence it must contain hydrogen and carbon, which are the main ingredients of coal-gas. This gas is in fact identical with coal-gas. The heat of the burning match converts some of the wax of the candle into inflammable gas. This gas takes fire, and in burning develops sufficient heat to convert more of the wax into gas, which is consumed as fast as formed.

192. *The Flame of a Candle consists of three distinct Parts.* — In the flame of a candle there are three distinct parts:

(1.) the dark central zone, or supply of unburnt gas surrounding the wick ; (2.) the luminous zone, or area of incomplete combustion ; and (3.) the non-luminous zone, or area of complete combustion. If we put one end of a small glass tube (see Figure 45) into the dark central zone, the unburnt gas will pass up the tube, and may be ignited at the other end. In the luminous part of the flame, as we have seen in the case of the gas-flame (186), the gas is not completely burnt, and carbon is separated in the solid state; and it is this carbon heated white-hot which renders the flame luminous. In the outer zone the supply of oxygen is greater, all the carbon is at once burnt to carbonic acid, and the flame here becomes non-luminous.

Fig. 45.

193. *The Blowpipe Flame consists of two Parts.* — A blowpipe is an instrument for sending a jet of air or oxygen into a flame, in order to make its heat more intense. In this way a common flame is converted into a kind of furnace, capable of fusing or raising to a high temperature any small body exposed to its action. The blowpipe, in one form or another, is very extensively used in chemistry and the arts.

The common mouth-blowpipe is merely a bent brass tube, tapering to a very fine jet.

The blowpipe flame consists of two distinct parts (see Figure 46): (1.) the *oxidizing* flame, *a*, where there is excess of oxygen; and (2.) the *reducing* flame, *d*, where there is excess of carbon. These correspond to the outer and middle zones of the candle-flame.

Fig. 46.

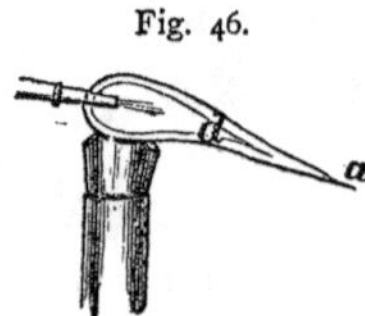

If the end of a copper wire be introduced into the oxidizing part of the flame, it at once combines with the heated oxygen, and becomes coated with oxide. If now it be put into the *reducing* flame, the heated carbon at once takes the oxygen away, and reduces it to the metallic state (140).

SUMMARY OF DESTRUCTIVE DISTILLATION AND OF ILLUMINATION.

Vegetable compounds may be broken up by means of heat into a great number of substances, which vary with the temperature to which the compounds are exposed.

This process is called *destructive distillation.* (163.)

The preparation of charcoal is one of the simplest cases of destructive distillation. In this case the wood is decomposed into carbon and a number of volatile compounds. (164.)

When these volatile products are cooled and condensed, they give rise to three classes of compounds: (1.) permanent gases; (2.) a watery fluid; (3.) a dark resinous fluid.

(1.) The permanent gases are chiefly two *hydrocarbons*, or compounds of carbon and hydrogen, called *marsh-gas*, H_4C, and *olefiant gas*, H_2C.

(2.) The chief ingredient of the watery fluid is *pyroligneous acid*, or *wood vinegar*, a compound of carbon, hydrogen, and oxygen, from which acetic acid is prepared.

(3.) The dark liquid is *wood tar*, which contains a variety of substances, mostly compounds of carbon and hydrogen.

The most important of these ingredients of wood tar are *wood naphtha*, *creosote*, *paraffine*, and *asphalt.* (165 – 169.)

Mineral coal has probably been formed by the gradual decomposition of vegetable substances buried in the earth, and thus excluded from the air. It is probable that this decomposition has, in many cases, been hastened by the internal heat of the earth. (171.)

When the volatile products were allowed to escape freely, *anthracite* coal was formed. When these products could not escape at the time of their formation, *cannel* and other varieties of *bituminous* coal were formed. (172.)

The volatile products often escaped into large cavities in the rocks, and on condensing gave rise to the celebrated *rock-oil*, or *petroleum.*

In some cases this oil may have been formed by the gradual decomposition of the vegetable substances at a temperature too low for its conversion into vapor, and may afterwards have been distilled by the internal heat of the earth. (179.)

The products of the destructive distillation of bituminous coal are even more numerous than those of the distillation of wood. These products are not so rich in compounds containing *oxygen*, such as acetic acid and creosote; but they are richer in compounds containing *nitrogen*, such as ammonia and aniline.

When the coal is distilled at a high temperature, a large amount of gas is obtained, and the liquid products are then called *tars.* When the coal is distilled at a low temperature, only a small amount of gas is obtained, and the liquid products are called *oils.* (173.)

When the gas obtained by the distillation of coal is purified, it forms the well-known *illuminating-gas.* (182.)

The most important ingredients of *coal tar* are *carbolic acid*, *ammonia*, *aniline*, *benzole*, *toluole*, *cumole*, *paraffine*, and *naphthaline.* (174–177.)

The celebrated *aniline colors* are prepared chiefly from benzole. (174, 175.)

When the coal is distilled at a high temperature, benzole and naphthaline are formed in much greater quantities than when it is distilled at a low temperature; while in the latter case toluole, cumole, and paraffine are more abundant products than in the former. (177.)

Solids burn without flame. Only gases burn with flame.

Solids that appear to burn with flame are first converted into a gas by heat, and this gas burns with flame. (183.)

The two conditions essential to illumination are the presence of a *solid* and intense *heat*. (185.)

The illuminating gases are all *hydrocarbons*, or compounds of hydrogen and carbon.

The hydrogen has the stronger affinity for oxygen, and burns first. The carbon is set free in a solid state, and, delaying for an instant in the intense heat of the flame, becomes luminous. It then combines with oxygen, forming an invisible gas, carbonic acid. (186.)

The light of the flame is increased by increasing the surface of the flame. Hence the ordinary gas flame is made flat, or, as in the case of the Argand burner, cylindrical. (188, 189.)

The light is also increased by increasing the intensity of the combustion of the hydrogen, as in the Bude light. (190.)

The heat of the flame is increased by mixing oxygen with the gas before it burns, so that the carbon and hydrogen may burn together, as in the Bunsen burner and in the blowpipe flame. (189, 193.)

The burning candle or oil lamp is a miniature gas-factory. (191.)

The candle flame consists of three parts: (1.) an inner zone of unburnt gases, surrounded by (2.) a luminous zone of incomplete combustion, outside of which is (3.) a non-luminous zone of complete combustion. (192.)

The blowpipe flame consists of two parts: (1.) the outer, or *oxidizing* flame, and (2.) the inner, or *reducing* flame. (193.)

THE ALKALIES AND THEIR MANUFACTURE.

194. *The Alkalies.* — The substances commonly known as the alkalies are *potash*, *soda*, and *ammonia*. They have been known from time immemorial. Potash and soda are solids, and were therefore called by the alchemists the *fixed* alkalies. Ammonia is a gas, and was therefore called the *volatile* alkali. Of all the bases, these three substances are endowed with the most powerful affinities; hence their importance in chemistry and the arts. Their salts are all soluble.

195. *Soap.* — If a weak solution of soda is boiled with beef-tallow for half an hour, and then a strong solution of soda is gradually added while the boiling continues, the mixture becomes by degrees of a uniform gluey consistency. If now some common salt be added, there is obtained a solid mass floating on a watery liquid, which holds the salt and some soda in solution.

The solid mass is common hard soap, which is insoluble in water in which salt is dissolved.

If this experiment is repeated, using palm or olive oil instead of tallow, hard soap is also obtained.

If, in any of these cases, potash is used instead of soda, soft soap is formed.

196. *Glycerine.* — The liquid obtained in the preparation of hard soap contains in solution not only salt and soda, but also a substance called *glycerine*. This substance may be prepared in a state of purity, by treating oil with oxide of lead instead of potash or soda. In this case, an insoluble lead soap is formed, while the glycerine remains liquid.

197. *Soaps, Oils, and Fats are Salts.* — Dissolve some soda soap in water, and to the solution add acetic acid

drop by drop, till it becomes turbid. A solid is formed, which is lighter than water, and at the same time insoluble in it, since it floats on the surface. This solid resembles tallow in appearance, but it has an acid reaction. The acetic acid sets this acid free from the soap, and itself combines with the soda. Soaps are thus seen to be made up of an acid and a base, and are true salts. Oils and fats are also salts. The base in all the oils and fats is glycerine. In tallow there are two acids, *stearic* and *oleic.* In oils there are acids analogous to these. The acid in palm-oil is called *palmic acid.*

The hard soaps, then, are *stearates*, *oleates*, *palmates*, etc. of *sodium;* and the soft soaps are the corresponding salts of *potassium.*

Rosin is also largely used in the manufacture of hard soap. It acts as an acid, and helps to saturate the soda.

Owing to its cleansing properties, soap is used in enormous quantities. In Great Britain alone, about 100,000 tons are manufactured annually.

In making this soap large quantities of the fixed alkalies are consumed.

198. *Potash.*—If a handful of wood-ashes is placed on a filter, and hot water gradually poured over it, the liquid filtered through will turn red litmus-paper blue. If this liquid is evaporated to dryness, a gray solid remains behind, which becomes white on being heated to redness. On holding some of this solid in the flame of a Bunsen's lamp, by means of a platinum wire, the flame becomes rose-colored, showing that the substance contains potassium (46). If any acid be added to this solid, carbonic acid escapes, showing that it is a carbonate. It must then be *carbonate of potassium*, $K\text{€}_3\Theta$. When prepared in this way, it is called *crude potash.* In regions where wood is plenty, as in Russia and America, crude potash is manufactured on a large scale by a process similar to this.

This method of separating soluble substances from insoluble ones is called *leaching*, or *lixiviation*.

There are in the ashes of plants other soluble substances besides carbonate of potassium, and all these will, of course, be found in crude potash.

If crude potash is dissolved in an equal weight of water, and allowed to stand over night, a sediment settles which is chiefly *silica*, or *silicic acid* (89). If the solution is then evaporated one half, and again allowed to stand over night, the other foreign substances will mostly crystallize. If it be then filtered, and the filtrate evaporated to dryness, *purified potash* is obtained.

Potash is very soluble. It is therefore the first substance to dissolve, and the last to separate from the solution. The other substances found in ashes are much less soluble ; hence they remain partially undissolved, and partially separate from the liquid in crystals, before the potash shows even the slightest tendency to crystallize.

If a solution of potash is boiled with hydrate of lime, the potassa and lime change places.

$$K_2\Theta,\ \text{Є}\Theta_2 + \text{Є}a\Theta,\ H_2\Theta = K_2\Theta,\ H_2\Theta + \text{Є}a\Theta,\ \text{Є}\Theta_2,$$

$$\text{or,}\quad 2K\text{Є}\Theta_3 + H_2\text{Є}a\Theta_2 = 2HK\Theta + \text{Є}a\,(\text{Є}\Theta_3)_2.$$

In this way, the hydrate of potassium, which is used in the making of soap, is prepared from the carbonate, which is the chief commercial salt.

199. *Soda-ash.* — While the ashes of land plants contain a large amount of potassa, those of marine plants contain little potassa, but a large amount of soda. It was from the ashes of such plants that carbonate of sodium, or the soda-ash of commerce, was formerly obtained.

Before the French Revolution, the Continental nations of Europe derived their soda-ash mainly from Spain, where it was made from the ashes of certain marine plants, which were extensively cultivated for this purpose. The soda-ash obtained from the ashes of these plants was called *barilla.*

In England, the soda-ash used was mainly obtained from the ashes of a sea-weed called *Kelp*, which grows abundantly on the north and west coast of Ireland, and on the west coast and the islands of Scotland.

One of the first effects of the war of the French Revolution was to cut off the supply of alkali from Spain. About this time, a French chemist, LeBlanc, discovered a process by which carbonate of sodium, or soda-ash, could be obtained from common salt, or *chloride of sodium*, NaCl. This process became publicly known through a commission appointed, during the first year of the Republic, to investigate the subject of alkali manufacture. As it met with their approval, it was soon carried into execution.

The method discovered by LeBlanc is the same as that now used in the making of carbonate of sodium, which has come to be one of the most important branches of chemical manufacture.

200. *The Manufacture of Carbonate of Sodium.* — The manufacture of carbonate of sodium from chloride of sodium may be divided into two stages : —

(1.) Manufacture of sulphate of sodium, or *salt-cake*, from chloride of sodium ; called the *salt-cake process.*

(2.) Manufacture of carbonate of sodium, or soda-ash, from salt-cake ; called the *soda-ash process.*

(1.) *Salt-cake Process.* — This consists in the decomposition of salt, by means of sulphuric acid. This is effected in a furnace, called the *salt-cake furnace*, a section of which is represented in Figure 47. It consists

Fig. 47.

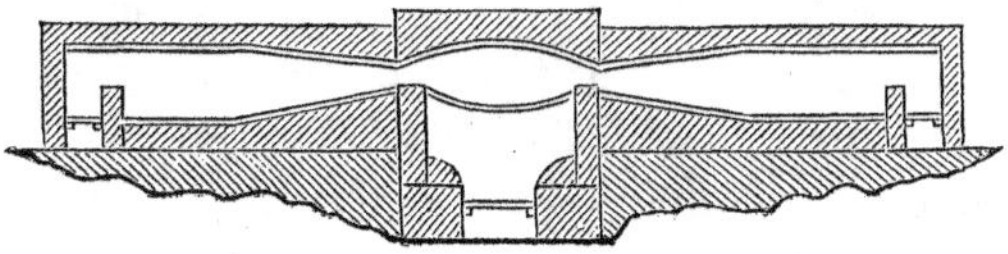

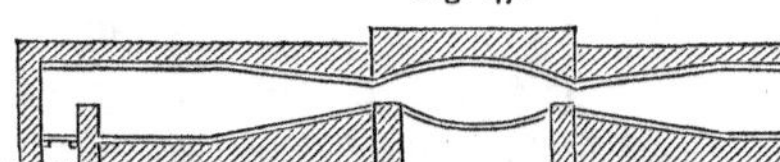

of a large covered iron pan, placed in the centre of the furnace, and heated by a fire placed underneath; and two *roasters*, or reverberatory furnaces (140), placed one at each side, on the hearths of which the salt is completely decomposed. The charge of half a ton of salt is first placed in the iron pan, and the requisite quantity of sulphuric acid allowed to run in upon it. Hydrochloric acid is evolved, and escapes through a flue with the products of combustion into towers, or *scrubbers*, filled with coke or bricks moistened with a stream of water; the acid vapors are thus condensed, while the smoke and heated air pass up the chimney. The reaction is as follows:—

$$2NaCl + H_2\Theta, S\Theta_3 = Na_2\Theta, S\Theta_3 + 2HCl.$$

After the mixture of salt and acid has been heated for some time in the iron pan, and has become solid, it is raked out upon the hearths of the *roasters*, where the flame and heated air of the fire complete the decomposition into sulphate of sodium and hydrochloric acid.

(2.) *Soda-ash Process.*—This consists (1.) in the preparation of the carbonate of sodium, and (2.) in the separation and purification of the same. The first chemical change which the salt-cake undergoes is its reduction to *sulphide* of sodium, by heating it with powdered coal:—

$$Na_2\Theta, S\Theta_3 + C_4 = Na_2S + 4C\Theta.$$

The second decomposition is the conversion of the sulphide of sodium into carbonate of sodium, by heating it with chalk or limestone (carbonate of calcium):—

$$Na_2S + Ca\Theta, C\Theta_2 = Na_2\Theta, C\Theta_2 + CaS.$$

These two reactions are, in practice, carried on at once; a mixture of ten parts of salt-cake, ten parts of chalk or limestone, and seven and a half parts of coal, being heated in a reverberatory furnace, called the *balling furnace*, until it fuses, and the decomposition is complete, when it is raked out into iron wheelbarrows to cool. This process is generally called the *black-ash process*, from the color of the fused mass.

The next operation consists in the separation of the carbonate of sodium from the insoluble sulphide of calcium and other impurities. This is easily accomplished by *lixiviation*, or dissolving the former salt out in water. On evaporating down the solution, for which the waste heat of the balling furnace is used, and calcining the residue, the soda-ash of commerce is obtained.

No less than 200,000 tons of common salt are annually consumed in the alkali works of Great Britain, for the preparation of nearly the same weight of soda-ash, the value of which is about £2,000,000.

Hydrate of sodium, or *caustic soda*, is made in large quantities by boiling soda-ash and hydrate of calcium together, and then evaporating the solution. The reaction is similar to that in the making of hydrate of potassium, or caustic potash (198).

201. *Bicarbonate of Soda.* — Bicarbonate of soda is obtained by exposing the crystallized carbonate of sodium in an atmosphere of carbonic acid. It is a white crystalline powder, chiefly used in medicine and for making effervescing drinks. It is also much used in bread-making as a substitute for yeast.

THE MANUFACTURE OF SULPHURIC ACID.

202. As soon as the manufacture of soda-ash from salt was established, the demand for sulphuric acid increased to an immense extent, and capital was freely invested to supply it. The origin and formation of sulphuric acid were carefully studied, and from year to year better, simpler, and cheaper methods of making it were discovered. With

every improvement in the mode of manufacture its price fell, and its sale was increased in an equal ratio.

Sulphuric acid was first prepared by distilling *ferrous sulphate*, or *green vitriol*, FeO, SO_3. The acid thus obtained is known as *fuming* or *Nordhausen* acid, and consists of a mixture of *sulphuric anhydride* and *hydrate of sulphuric acid*, $SO_3 + H_2SO_4$ (94–96).

This method has long been superseded by a more convenient one, which depends upon the fact that, though sulphurous acid, SO_2, does not combine with free oxygen and water to form hydrate of sulphuric acid, it is capable of taking up the oxygen when the latter is united with nitrogen in the form of nitrous acid, N_2O_3 (81, 82). Thus:

$$SO_2 + 2H_2O + N_2O_3 = H_2SO_4 + 2NO.$$

Sulphurous acid, water, and nitrous acid yield sulphuric acid and nitric oxide.

The nitric oxide formed in this decomposition takes up another atom of oxygen from the air, becoming nitrous acid, N_2O_3, and this is again able to convert a second molecule of sulphurous acid, SO_2, with water, H_2O, into sulphuric acid, H_2SO_4; being a second time reduced to nitric oxide, NO, and ready again to take up another atom of oxygen from the air. Hence it is clear that the nitric oxide acts simply as a carrier of oxygen between the air and the sulphurous acid; and a very small quantity of it is therefore able to convert a very large quantity of sulphurous acid, water, and nitrous acid into sulphuric acid.

This process is conducted on a large scale in chambers made of sheet-lead, which are often of a capacity of 50,000 or 100,000 cubic feet. The sulphurous acid is obtained either by burning sulpnur in a current of air, or by roasting a mineral called *iron pyrites* (a *sulphide of iron*, FeS_2), and is led, together with air, into the chamber. The nitrous acid is got from nitrate of sodium (103), which is decomposed either by the heat of the burning sulphur, or in a separate furnace. Jets of steam are also blown into the chamber at various points, and a thorough draft is maintained by a high chimney. The sulphuric acid, as it forms, falls to the floor of the chamber, whence it is continually drawn off. It is then heated, first in open leaden pans, and then in vessels of glass or platinum (as lead is attacked by the strong acid), until the excess of water is driven off, and a sufficiently concentrated acid is obtained.

So extensively is this acid used, that the quantity manufactured in South Lancashire alone exceeds 3,000 tons a week.

THE MANUFACTURE OF BLEACHING POWDER.

203. *The Bleaching Power of Chlorine.* — We have seen that in the manufacture of salt-cake hydrochloric acid is an incidental product: —

$$2NaCl + H_2SO_4 = Na_2SO_4 + 2HCl.$$

This hydrochloric acid was at first allowed to escape by the chimney as worthless, but a profitable application of it was soon discovered.

The bleaching power of chlorine has long been known; but it came to be employed on a large scale only after it was obtained from this waste muriatic acid.

We have seen (61) that if a piece of litmus-paper be introduced into chlorine gas, it becomes white. We have seen also that chlorine is readily absorbed by water, which is then called *chlorine water.* Pour chlorine water upon wine or red ink, and the liquids will lose their color.

Chlorine bleaches and destroys all organic colors; that is, all colors derived from the vegetable and animal kingdoms.

In consequence of this property, and of the cheapness with which it can be prepared from the waste hydrochloric acid mentioned above, chlorine has become an important agent in the art of bleaching. It is used chiefly in bleaching cotton fabrics, which may now be rendered perfectly white in a few hours; while, by the old method of laying them on the grass in the sun, it required weeks, and even months, to effect it. It was necessary, moreover, to have meadow-land suitably situated for the bleaching; hence most of the cloth manufactured in England was carried to Holland to be bleached. Besides the diminished expense, the cotton stuffs bleached with chlorine suffer less injury in the hands of skilful workmen than those bleached in the sun.

204. *Bleaching Powders.* — The health of the workmen is greatly endangered by the use of chlorine in the gaseous or liquid state; but it has been found that the gas is readily absorbed by slaked lime, and is as readily given up again when the lime is treated with dilute acid.

When thus combined with lime, the chlorine can be easily transported. The compound is called *chloride of lime*, or *bleaching powder.*

205. *The Preparation and Use of Bleaching Powder.* — This substance is made on a large scale by conducting chlorine into spacious chambers, on the floor of which slaked lime is spread to the depth of two inches.

The chlorine is prepared by heating a mixture of hydrochloric acid and black oxide of manganese, MnO_2: —

$$MnO_2 + 4HCl = MnCl_2 + 2H_2O + 2Cl.$$

Black oxide of manganese is found in many parts of Europe, but the most productive mines are in Thuringia and Moravia. More than 18,000 tons are used annually in England in the manufacture of bleaching powder.

In bleaching, the goods are first dipped into a solution of the bleaching powder, and then passed through dilute acid. The chlorine is thus set free in the fibres of the cloth.

PRODUCTS OF THE SODA-ASH MANUFACTURE, AND THE MATERIALS USED.

206. The direct and indirect products of the soda manufacture are chiefly carbonate of sodium, bicarbonate of soda, hydrate of sodium (caustic soda), sulphuric acid, hydrochloric acid, and bleaching powder. The raw materials used are, mainly, common salt, sulphur, nitrate of sodium (cubic nitre, or soda saltpetre), black oxide of manganese, and lime.

SOURCES OF SALT.

207. *Salt from Sea-Water.* — In Southern Europe, on the shores of the Mediterranean, large quantities of salt are obtained from sea-water. The water is allowed to flow into large shallow pools, called *salt-pans*, where it is evaporated by the agency of the air and the sun. Only pure water passes off in the form of vapor, and the solution grows more and more concentrated. After a time the brine is pumped into large iron pans, and evaporated by artificial heat until the salt crystallizes.

208. *Salt Lakes.* — In many parts of the earth there are salt lakes. The streams which flow into them are more or less impregnated with salt, which they dissolve from the soil over which they flow. Such lakes are natural salt-pans, for the water which is evaporated from them is pure, and they consequently become more and more saturated with salt till it is finally deposited on the bottom in solid crystals. There are many inland lakes in which salt is thus accumulating. The most remarkable of them is the Great Salt Lake in Utah.

209. *Rock Salt.* — Such salt lakes appear to have existed in all geological ages, and by long-continued evaporation to

have given rise to large masses of solid salt, called *rock salt.* These beds of rock salt are found in all parts of the earth. The most remarkable are at Wieliczka in Poland, and at Cardona in Spain. The former is 500 miles long, 20 miles broad, and 1200 feet deep. It has been worked for several centuries. Some of the galleries excavated in it are 30 miles long. The bed at Cardona is a "mountain of salt," 400 or 500 feet high, and the salt is of the greatest purity.

Salt is obtained from these beds sometimes by excavation, and sometimes by solution. In the latter case, a hole is sometimes bored down to the bed of salt, often to the depth of 1200 or 1500 feet. A tube somewhat smaller than the bore is then introduced, and fresh water allowed to flow down outside the tube. This water, on reaching the bed, becomes saturated with salt, and rises in the tube. Since, however, it is heavier than fresh water, it will not rise to a level with the latter on the outside of the tube. If the bore is 1200 feet deep, the brine will rise within about 200 feet of the top. It must be pumped up the remaining distance. The solution is then evaporated, and solid salt obtained.

The arrangement just described may be regarded as an artificial spring.

210. *Salt Springs.* — It not unfrequently happens that natural springs occur in such localities that their waters come into contact with beds of salt. Such springs are called *salt springs*, or *brine springs.* They sometimes come to the surface; but oftener they can only be reached by boring, and the brine brought to the surface by means of the pump. The water of these natural springs is usually less concentrated than that of the artificial springs. In England and in this country the salt produced is nearly all obtained from brine springs.

The most important salt springs of this country are in Central New York, in Virginia, and in Pennsylvania.

About 6,000,000 bushels of salt are produced annually in New York. When the brine is not a saturated solution, it is usually concentrated by partially evaporating it in large shallow pans by the air and the sun. It is then further concentrated and crystallized by artificial heat.

SOURCES OF SULPHUR.

211. Sulphur occurs in nature both free and combined: it is found free in certain volcanic countries, especially in Sicily and Iceland; it exists in combination with many metals, forming *sulphides* (138, 140). These sulphides are the ores from which several of the metals are commonly obtained. Thus PbS, sulphide of lead (galena), ZnS, sulphide of zinc (zinc blende), and CuS, sulphide of copper, are the most productive ores of those metals.

In order to obtain pure sulphur, the mineral containing the crude substance mixed with earthy impurities is heated in earthen pots (Figure 48); the sulphur distils over in

Fig. 48.

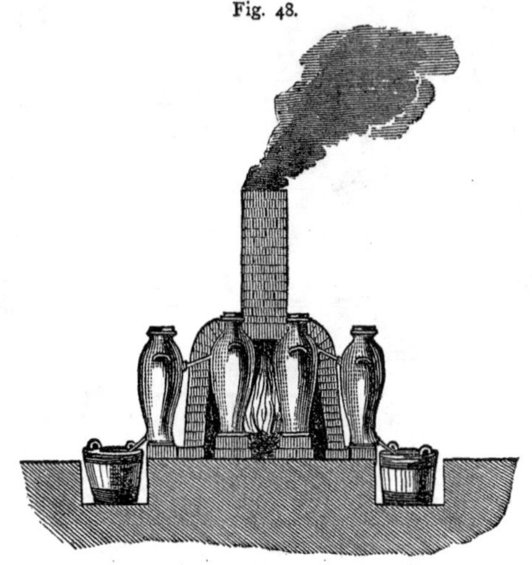

the form of vapor, which is condensed in similar pots placed outside the furnace. The sulphur thus obtained is refined or purified by a second distillation.

If the vapor of sulphur is quickly cooled below its melting-point, it solidifies in the form of a fine crystalline powder, called *flowers of sulphur;* just as watery vapor, when cooled below the freezing point of water, becomes snow. When sulphur is gently heated it melts, and may be cast into sticks, and is then known as *brimstone* or *roll sulphur.*

SOURCES OF AMMONIA.

212. Ammonia is usually obtained by the action of lime on sal ammoniac. The lime should be previously slaked, and then mixed with about two thirds of its weight of finely powdered sal ammoniac. The mixture is heated in a flask arranged as shown in Figure 49. The gas is washed by passing it through a wash-bottle, and is then conducted into water. This solution, as we have already seen (43), is known by the name of *aqua ammonia.* At the ordinary temperature, water absorbs more than 1,000 times its bulk of the gas.

Fig. 49.

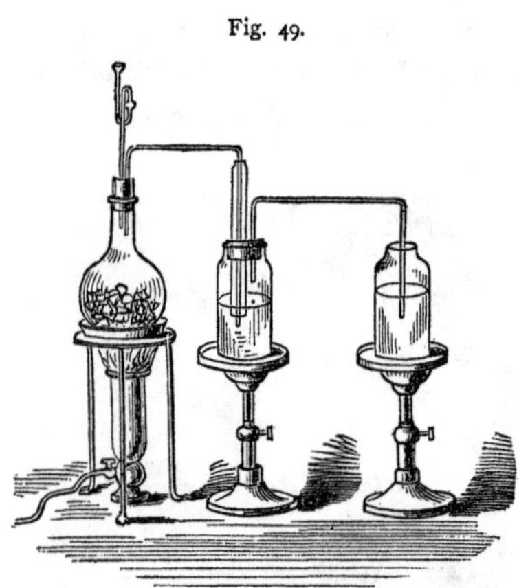

The reaction between hydrate of lime and sal ammoniac is as follows :—

$$2(H_4N)Cl + H_2CaO_2 = CaCl_2 + 2H_2O + 2H_3N.$$

Sal ammoniac, or *chloride of ammonium* (100), is obtained from the *ammoniacal liquor* of the gas-works (182), by treating it with hydrochloric acid.

SUMMARY OF THE MANUFACTURE OF THE ALKALIES.

The three most important alkalies are the hydrates of potassium, sodium, and ammonia. The first two are called the *fixed* alkalies; the last, the *volatile* alkali. (194.)

The fixed alkalies are usually prepared as carbonates, and are rendered caustic by the addition of caustic lime. The carbonate of potassium is commonly known in commerce as *potash*, and the carbonate of sodium as *soda-ash.* (198, 199.)

These alkalies are used in enormous quantities in the manufacture of *soap.* (195.)

The hard soaps are *stearates*, *oleates*, *palmates*, etc. of *sodium*, and the soft soaps are the corresponding salts of *potassium.* (197.)

Potash is obtained from the ashes of land plants by *lixiviation.* (198.)

Soda-ash was formerly obtained from the ashes of marine plants by the same method. It is now made in enormous quantities from common salt, or chloride of sodium.

The chloride of sodium is first treated with hydrate of sulphuric acid, which gives rise to sulphate of sodium and muriatic acid.

This sulphate is then converted into carbonate by heating it with coal and chalk. The carbonate of sodium is then separated by lixiviation from the sulphide of calcium formed at the same time, and is crystallized by evaporating the solution thus formed. (199, 200.)

Large quantities of caustic soda and of bicarbonate of soda are made from a portion of this soda-ash, at the manufactories of the latter. (200, 201.)

The soda-ash is converted into bicarbonate of soda, by exposing the crystals to the action of carbonic acid. (201.)

The muriatic acid produced in the manufacture of soda-ash is saved, and used as a source of chlorine for the making of bleaching powder.* (200, 203.)

The chlorine is separated from the muriatic acid by the action of black oxide of manganese, and bleaching powder is formed by the action of this chlorine on slaked lime. (205.)

Great quantities of sulphuric acid are consumed in the manufacture of soda-ash.

This acid, in a dilute state, is formed by the mutual reaction of sulphurous acid, nitric oxide, air, and steam, in leaden chambers, and is concentrated by boiling it down in glass or platinum retorts.

The sulphurous acid is obtained by burning sulphur in a free current of air, and the nitric oxide by the action of sulphuric acid on nitrate of sodium. (202.)

The sulphur used comes mainly from Sicily, and the nitrate of sodium from Peru. (211, 161.)

Chloride of sodium, or salt, is obtained in various parts of the earth from salt beds and springs. (209, 210.)

Ammonia is prepared from the ammoniacal liquor, which is a waste product of the gas-works. (212.)

* See Appendix, 30.

CONCLUSION.

We have now examined a few chemical substances and processes, and have thus become familiar with the action of the force called *affinity*.

An examination of the four compounds muriatic acid, water, ammonia, and marsh-gas, and of the compounds of nitrogen and oxygen, brought clearly to light the leading features of the action of this force.

In the first place, we learned that affinity causes the atoms of different elements to unite, so as to form new molecules with properties quite unlike the original ones. Secondly, we found that it always causes a definite number of atoms to unite to form a molecule of the new compound. Thirdly, it acts between the atoms of different elements with different degrees of strength. Fourthly, it acts by preference between one atom of a given element, and a certain number of atoms of other elements, which is the same for elements of the same group, and different for those of different groups. Fifthly, it often causes the atoms of two or more elements to unite in several proportions to form different compounds, which have quite different properties.

Then, in examining the various ways in which binary and ternary salts are formed, we discovered that when two compounds are brought together under favorable circumstances, the atoms of the different elements often change places. When the atoms belong to the same group, an atom of each element replaces an atom of another. When they belong to different groups, an atom of one no longer replaces an atom of another; but they replace one another in such a way that the atom-fixing power of the atoms in a molecule of each of the new compounds remains the same as in the old ones.

Next, an examination of ordinary combustion revealed the fact that the burning of wood and coal, and of gas, are chemical processes; and that our artificial light and heat are developed by the action of affinity; while it brought clearly to light the fact, that affinity is often dormant till it is roused to activity by some agent, such as heat. The study of this process also disclosed the part that affinity had played in the formation of our earth.

The study of respiration, decay, and the growth of plants showed the part which affinity now plays in nature. We found that the atmosphere is transformed into food fit for man and many other animals by a chemical process carried on in the leaf by the agency of the sunbeam. The enormous amount of material thus removed from the atmosphere we found to be again poured back into it as the products of another chemical process, by which organic substances are destroyed in ordinary combustion, respiration, and decay. And we found that this chemical process is maintained by the oxygen set free in the growth of plants.

We noticed also, that the chemical process going on in the leaf furnishes us, not only with our food and with fuel for our fires, but with the means of again liberating the most useful metals from the elements with which affinity had locked them up in the formation of the present materials of our globe.

A brief study of destructive distillation brought to our view a few of the most important of the substances which can be produced by the breaking up of vegetable compounds, by means of heat. Among these substances, we noticed particularly the various kinds of coal, illuminating-gas, coal-oils, paraffine, ammonia, and aniline.

The substances that can be obtained by the breaking up

of vegetable compounds by the action of heat and chemical agents are almost literally innumerable, and they are daily put to new and important uses. But a short time ago coal-tar, one of the products of the destructive distillation of vegetable matter, was regarded as worse than useless. From one of its ingredients there is now prepared a whole series of the most brilliant and valuable dyes. This most important field of chemistry has been but slightly explored as yet, and only a very small portion of it can be said to be accurately described.

The study of illumination revealed one of those nice adjustments of affinity, by which provision has been made in nature for the wants of man.

The study of the manufacture of the fixed alkalies then illustrated the action of affinity in one of the most important branches of chemical manufacture, and the improvements made in this manufacture through a better knowledge of that action. At first, we find soda-ash prepared by lixiviation of the ashes of marine plants. Then, a better knowledge of chemical action led to the discovery that the sodium of common salt could be converted into soda-ash. But this process required a large amount of sulphuric acid. This acid had, heretofore, been made by the rude process of distilling green vitriol ; but the increased demand for it led to a more careful study of the action of affinity in its formation, and the important discovery was made that sulphurous acid (which can be produced in any quantity by simply burning sulphur in a free supply of air) can be readily converted into sulphuric acid, on a large scale, by means of nitric oxide and steam.

Until quite recently, all the hydrochloric acid produced in the manufacture of soda-ash was allowed to escape as a waste product, although the bleaching power of chlorine was well known. But a more thorough knowledge of affinity showed how this element could be safely and economically used for bleaching cotton fabrics. Then a way was found of saving the hydrochloric acid, and its chlorine was consumed in the preparation of bleaching powder.

We have not aimed to exhaust the subject of chemical affinity. We have merely studied a few compounds to discover the laws of its action, and a few processes to illustrate the importance of this force in nature and in the arts.

IV.

ELECTRICITY.

ELECTRICITY.

MAGNETISM.

213. *Magnets.*—In studying Electricity it is necessary to know a few things about *magnets.*

Bring one end of an ordinary bar magnet into contact with a pile of iron tacks; on removing it, a number of the tacks are carried away with it. This illustrates one of the leading characteristics of a magnet; namely, that it attracts iron. The force residing in a magnet and shown by its attracting iron is called *magnetism.*

There is a certain iron ore which has the power of attracting iron. This ore seems to have been first found near Magnesia, a city of Asia Minor. Hence the name *magnet.* Natural magnets are called *loadstones* (more properly *lodestones*), that is, stones that *lead* or draw iron.

214. *The Power of a Magnet resides chiefly at the Ends.*—If a small iron ball is fastened to a string, and moved alongside a bar magnet, it is scarcely attracted at the middle of the bar. As it approaches either end it is attracted more and more strongly, until it is brought near the end, where attraction is found much the strongest. The force of a magnet, then, resides chiefly at the ends.

Put a piece of stiff drawing-paper over a strong magnetic bar, and strew fine iron-filings over it. Not only is the position of the magnet below shown on the paper, but the particles of iron arrange themselves in lines radiating from the poles. These lines are called *lines of magnetic force*, or *magnetic curves.*

Fig. 50.

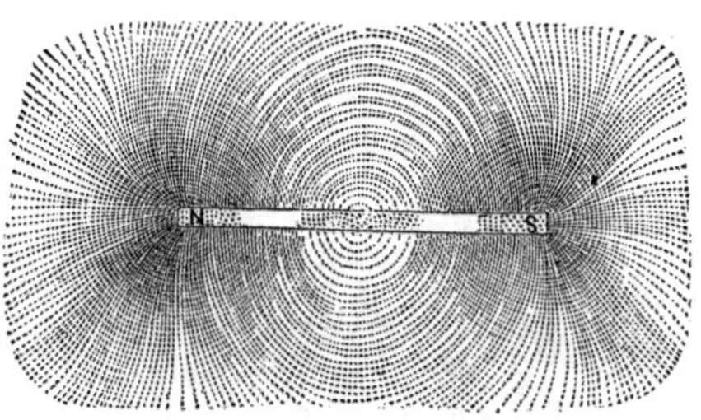

215. *The Forces at the Opposite Ends of a Magnet act in Opposite Directions.* — Suspend a bar magnet by a string so that it can turn freely. Bring one end of a bar magnet near one end of the suspended magnet, and the latter is drawn towards it. Reverse the ends of the bar magnet, and the end of the suspended magnet is repelled. This shows that the forces at the ends of a magnet act in opposite directions.

Fig. 51.

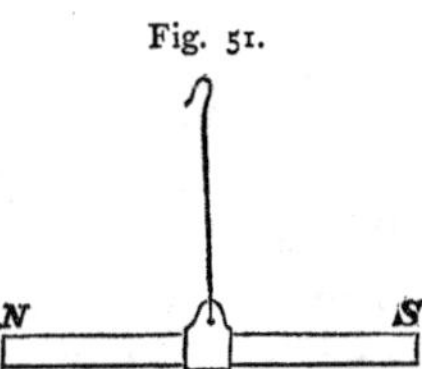

216. *The Poles of the Magnet.* — The ends of the magnet, where the opposite forces reside, are called *poles.*

When a bar magnet is poised so that it can move freely, it takes a nearly north and south direction. One of its poles will always point to the north, and is called the *north pole.* The opposite pole is called the *south pole.* A bar magnet thus poised so as to turn freely is called a *magnetic needle.*

217. *The Earth acts like a Magnet.* — If a small needle which is free to move in a horizontal plane is placed upon a bar magnet, its south pole will always point towards the north pole of the latter. If a small *dipping needle*, that is, a

needle which is free to move in a vertical plane, is placed above the middle of a bar magnet, it stands parallel with the bar magnet. If it is moved towards the north pole of the magnet, the south pole begins to dip towards the magnet; and the farther it is moved towards this pole, the more it dips. If it is moved from the centre of the bar magnet towards the south pole, its north pole dips in the same way.

We have already seen that a magnetic needle free to move in a horizontal plane points north and south when held above the earth. It is also found that a dipping needle in the vicinity of the equator stands parallel with the plane of the horizon, and that, when carried north from the equator, its north pole dips towards the horizon; while, if it is carried south from the equator, its south pole dips towards the horizon.

It is thus found that the earth acts upon a magnetic needle like a magnet whose poles are near the poles of the earth; its south pole near the north pole of the earth, and its north pole near the south pole of the earth.

The French regard the magnetic pole of the earth near the north pole as a north pole, and the pole of the magnetic needle which points towards this pole as the south pole. The pole of the magnet, therefore, which we call the north pole, the French call the south pole, and *vice versa.*

218. *Like Poles of Magnets repel and unlike Poles attract.* — This has already been shown by the action of a bar magnet upon a dipping needle. It may be further shown by bringing the north pole of a bar magnet near the north pole of a needle, which will be repelled. On bringing the south pole of this magnet near the north pole of the needle, it will be attracted.

219. *Magnetism is developed in Iron or Steel by Induction.* — When a piece of soft iron is brought into contact with the pole of a magnet, it will attract other pieces of

iron, showing that magnetism has been developed in the iron by contact with the magnet. Magnetism can be developed in a piece of steel in the same way. The iron however, loses its magnetism as soon as it is taken away from the magnet, while the steel retains it. It is not necessary that a piece of iron should be brought into actual contact with the pole of a magnet in order that magnetism may be developed in it, but merely that it be brought very near the pole.

Magnetism developed in this way in a piece of iron or steel is said to be *induced.*

220. *Forms of Magnets.* — Ordinary magnets are made of steel. When straight they are called *bar magnets;* when bent into the shape of the letter U they are called *horseshoe magnets.* When several bar or horseshoe magnets (see Figure 52) are connected, they constitute a *magnetic battery.*

Fig. 52.

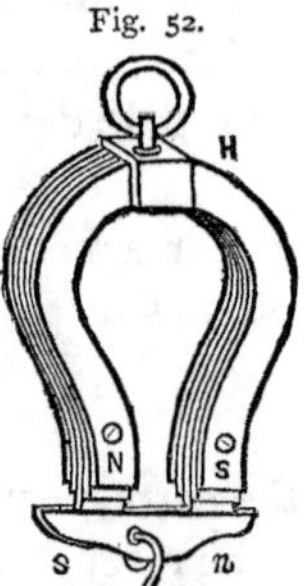

221. *The Making of Magnets.* — Magnets are often made by contact with permanent magnets by a process of *single* or *double touch.* In the former case, the steel bar to be magnetized is laid on a table, and the pole of a powerful magnet is rubbed from ten to twenty times along its length, always in the same direction. If the magnetizing pole be north, the end of the bar it first touches each time becomes north, and the end where it is taken off becomes south. The same thing may be done by putting one pole of the magnet, say the north, first on the middle of the bar, then giving it a few passes from the middle to the end, returning always in an arch from the end to the middle. The other half of the bar is then rubbed in the same way with the south pole of the magnet. The first end rubbed becomes the south, and the other the north pole of the new magnet.

The method by *double touch* is shown in Figure 53. The bar *s n* to be magnetized is placed on a piece of wood, *W*, with its ends resting on the extremities of two powerful magnets *NS* and *SN*. Two rubbing magnets are placed with their poles near, but not touching, on the middle of *s n*, inclined to it at an angle of 10° or 15°. The two magnets are then drawn along from the middle to one end and then back to the other, and so backwards and forwards from ten to twenty times, and lifted from the magnetized bar again at the middle. Care must be taken that both ends are rubbed the same number of times, and that the lower poles of the rubbing magnets do not go beyond the ends of the bar. Both the upper and lower surfaces of the bar must be rubbed in this way, in order to magnetize it fully. A small piece of wood may be placed between the poles of the rubbing magnets to prevent contact. The position of the poles is shown in the figure by the letters; *N* or *n* meaning a north, and *S* or *s* a south pole.

Fig. 53.

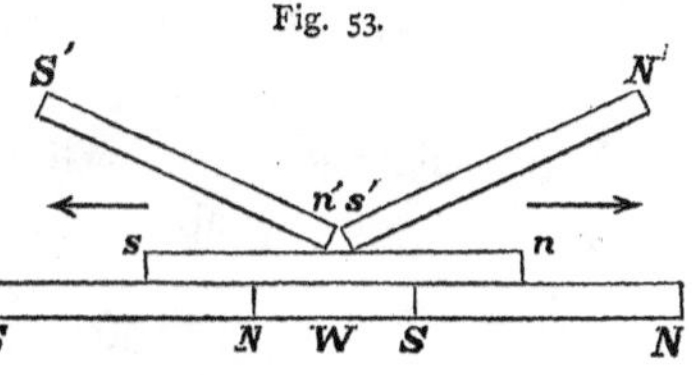

Fig. 54.

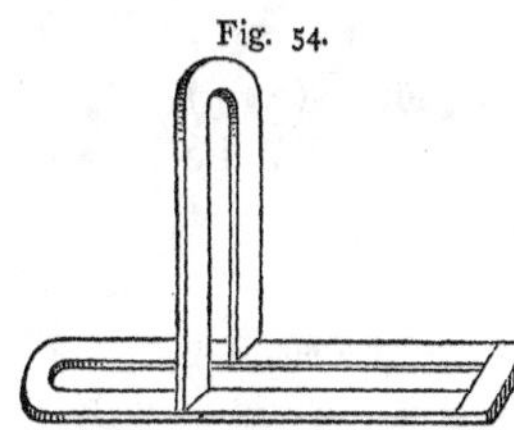

For horseshoe magnets Hoffer's method is generally followed. The inducing magnet (see Figure 54) is placed vertically on the magnet to be formed, and moved from the ends to the bend, or in the opposite way, and brought round again in an arch to the starting-point. A piece of soft iron is placed at the poles of the induced magnet. Both magnets should be of the same width.

SUMMARY.

Any substance which will attract iron is called a *magnet.* The force which enables it to attract iron is called *magnetism.* (213.)

This force resides chiefly at the ends of a magnet, which are called its *poles.* It radiates from these poles in curved lines, called *lines of magnetic force*, or *magnetic curves.* (214, 216.)

The forces residing at the opposite poles of a magnet act in opposite directions. (215.)

The north pole of a dipping needle always points towards the south pole of a bar magnet, when held over it.

The earth acts upon a needle like a magnet. Its magnetic poles are situated near the poles of its axis. As we name the poles of a magnet, the magnetic pole of the earth north of the equator is a *south* pole, and the one south of the equator is a *north* pole. The French call the magnetic pole north of the equator a *north* pole, and the end of the needle which points towards it a *south* pole. (217.)

Like poles of magnets repel, and unlike poles attract each other. (218.)

A magnet can develop magnetism in iron or steel by *induction.* Soft iron loses its magnetism as soon as it is withdrawn from the influence of the magnet, while steel retains its magnetism permanently. (219.)

Ordinary magnets are made of steel. They are called, from their shape, *bar magnets*, or *horseshoe magnets.* (220.)

Magnets may be made by contact with other magnets, by a process of *single* or *double touch.* (221.)

THE QUANTITY AND THE INTENSITY OF THE ELECTRIC FORCE.

222. The apparatus represented in Figure 55 is called a Bunsen's *cell* or *element.* It consists of the following parts: a large earthen or glass cup; a piece of zinc rolled into a cylinder and open down one side; a porous porcelain cup, small enough to go inside the zinc cylinder; and a piece of coke carbon small enough to stand inside the porcelain cup. These pieces are placed inside each other in the order mentioned; and the larger cup is filled with dilute sulphuric acid, and the smaller cup with the strongest nitric acid.

Fig. 55.

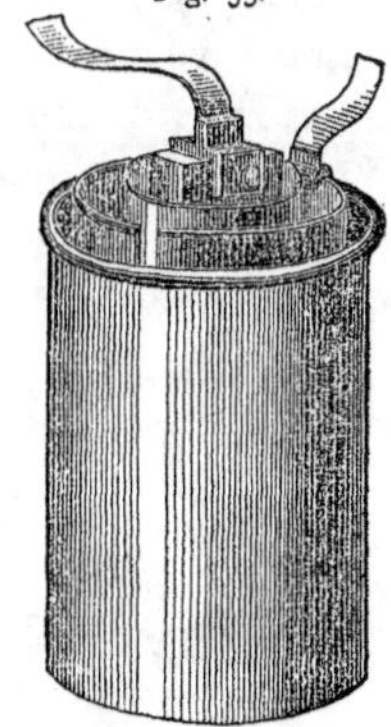

If now the zinc and carbon be connected by means of a copper wire, and this wire be held over a magnetic needle in such a way that the part of the wire above the needle shall be parallel with it, the needle at once turns aside; showing that there is a new force in the wire. This force is called *electricity.*

The parts of the cell thus connected are called the *poles.* In this case, the zinc is called the *negative* pole and the carbon the *positive* pole. Whatever connects the poles is called the *circuit.*

223. *The Electric Current.* — If the wire is held over the needle in such a way that the wire passes from the carbon of the cell to the north end of the needle, thence to the south end and to the zinc, the north pole of the needle turns to the east; or, supposing an observer to be facing the north pole of the needle, it turns to his left. If the direction of the wire is reversed, so that it passes from the

zinc to the north pole of the needle, thence to the south pole and the carbon, the north pole of the needle turns to the right.

The fact that the needle turns in opposite directions when the direction of the wire over it is reversed, suggests at once the idea that the force in the wire is in motion, or, in other words, is a *current*, which ought to produce opposite effects when flowing in opposite directions. It is called a *current of electricity*. As a matter of convenience, the electricity is considered as starting from the carbon, or the positive pole of the cell, and flowing through the wire to the zinc, or the negative pole.

It must be carefully borne in mind that by the electric current we mean simply the onward propagation of the electric *force*, and not the movement of any material substance or ethereal fluid. This flow of force from molecule to molecule is best illustrated by a set of ivory balls hung side by side, as shown in Figure 56. If the first ball at the left be raised and let fall again, all the balls remain at rest except the last which flies off. Obviously the force imparted to the first has passed from ball to ball throughout the line.

Fig. 56.

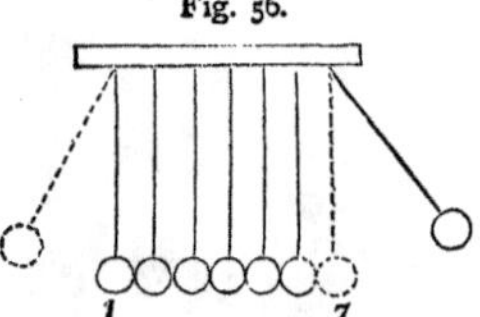

224. *The Electric Battery.*—When several cells are joined together, they constitute a *battery*. There are two ways in which the cells may be joined: (1.) the zinc of the first cell may be joined to the carbon of the second, and the zinc of the second to the carbon of the third, and so on throughout, and the free carbon, or positive pole, of the first cell joined to the free zinc, or negative pole, of the last, by means of a wire; or (2.) the zincs may all be joined together, and the carbons all joined together, and then the zincs and the carbons joined by a wire. These two ways of arranging the battery are shown in Figures 57 and 58.

Fig. 57.

Fig. 58.

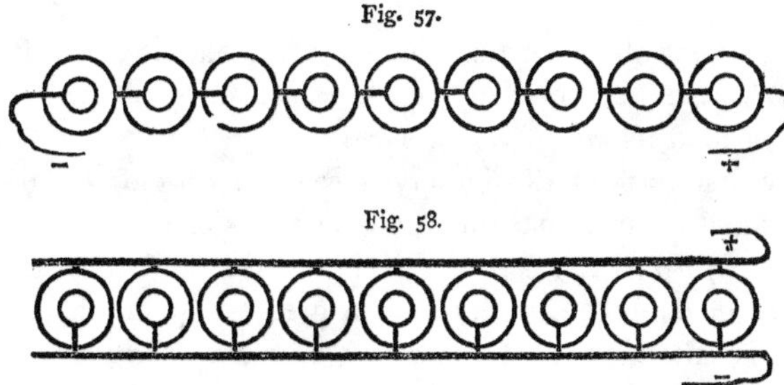

225. *Quantity and Intensity.* — Let a battery of three cells be arranged according to the first method, and the opposite ends, or *poles*, be connected by a short, thick copper wire, which passes over a needle, and let the number of degrees the needle is deflected be noted. Then let the same cells be arranged into a battery according to the second method, and the zinc and carbon be connected by the same wire, and the needle will be deflected more than in the former case. When the battery, then, is arranged according to the second method, the electricity which it generates has a greater power to deflect the needle than when it is arranged according to the first method.

If now the cells be again arranged as at first, and a piece of fine steel wire, five or six inches long, be introduced into the circuit, the needle will be deflected considerably less than when the circuit is completed with the short, thick copper wire; showing that the current is resisted in passing through the fine wire. If now the same piece of fine wire be introduced into the circuit, and the battery be arranged according to the second method, the needle will be deflected considerably less than when the battery is arranged according to the first method; showing that the electricity generated by the first form of battery has a greater power of overcoming resistance in the circuit than that generated by the second form.

The power of the current to deflect a needle is called its *quantity*, and its power to overcome resistance in the circuit is called its *intensity*, or its *tension*.

The first form of the battery is seen to develop electricity of the greatest tension, and hence it is called a *battery of tension*, or *intensity battery;* while the second form of the battery develops electricity in the greatest quantity, and hence is called a *battery of quantity*, or *quantity battery*.

When, therefore, much resistance is to be overcome, the tension battery had better be used; when but little resistance is to be overcome, the quantity battery is to be preferred.

Fig. 59.

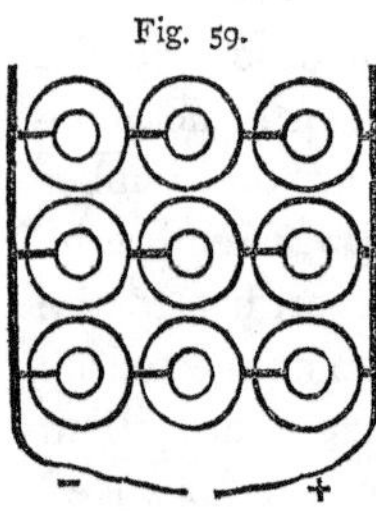

When a considerable quantity of electricity is required, and at the same time electricity of considerable tension, the two methods of arranging the battery are combined, as shown in Figure 59.

226. *Conductors and Non-Conductors.* — If a piece of glass, sealing-wax, or dry wood be introduced into the circuit, the needle will show that no current passes. Substances which will not allow the electric force to pass through them are called *non-conductors;* while other substances are called *conductors*. Metals are generally good conductors. Copper is one of the best conductors; hence copper wires are used in all the ordinary experiments with electric currents.

When the circuit is composed entirely of conductors, it is called a *closed* circuit; when there is a non-conductor in any part of the circuit, it is called an *open* circuit.

SUMMARY.

A Bunsen cell consists of a piece of coke carbon plunged in strong nitric acid contained in a porous cup, which is placed inside of a cylinder of zinc, plunged in dilute sulphuric acid.

The carbon and the zinc are called the *poles* of the cell; the carbon the *positive* pole, and the zinc the *negative* pole. That which is employed to connect the poles is called the *circuit.*

A force called *electricity* resides in a wire which connects the poles. (222.)

Since this force seems to flow through the wire, it is called a *current.* We always consider the current as starting from the positive pole, and passing to the negative pole. (223.)

When two or more cells are connected, the apparatus is called a *battery.*

The carbon of one cell may be connected with the zinc of the next, and so on throughout; or the carbons may all be connected, and also all the zincs. In the first case, the free zinc of the first cell and the free carbon of the last are the poles of the battery; in the second, the united carbons constitute one pole, and the united zincs the other. (224.)

When fine steel wire is introduced into the circuit, the current meets with resistance.

The power of the current to turn a needle is called its *quantity*, and its power of overcoming resistance is called its *intensity.*

The electric force does not always possess these two qualities in the same proportions.

The battery arranged in the first form gives electricity of

greater *intensity* than the battery arranged in the second form; while the latter gives electricity of greater *quantity* than the other. Hence the first form of battery is called an *intensity* battery; the second, a *quantity* battery. (225.)

Substances which allow the current to pass readily through them are called *conductors;* while those which will not allow it to pass are called *non-conductors.* The metals are generally good conductors, while dry wood, glass, resin, and gases are non-conductors. Copper is one of the best conductors.

When the circuit is made up entirely of conductors, it is said to be *closed;* when there is a non-conductor anywhere in the circuit, it is said to be *open.* (226.)

ACTION OF THE CURRENT UPON A NEEDLE.

227. *The Rheotome and the Rheotrope.* — In studying the action of the current upon a magnetic needle, it is desirable to have some ready way of stopping or breaking the current, and of changing its direction. An instrument for breaking the current is called a *rheotome,* a name derived from two Greek words, and signifying *current-cutter.* An instrument for changing the direction of the current is called a *rheotrope,* that is, a *current-turner.*

These two instruments are often combined in one. One of the most convenient forms of this apparatus is shown in Figures 60 and 61.

It has two binding-screws, by which it may be connected with the two poles of the battery. These binding-screws are connected with copper springs, which press against a key. *R* represents one of these springs, which are shown better at *m* and *n* in the end view given on a smaller scale in Figure 61. A copper strip runs from each end of the key, each strip terminating in a binding-screw, to which is attached the remainder of the circuit. The key (shown in section in Figure 60) is the essential part of the apparatus. It is made chiefly of ebony, which is a non-conductor of electricity. A metallic axis passes a little

Fig. 60.

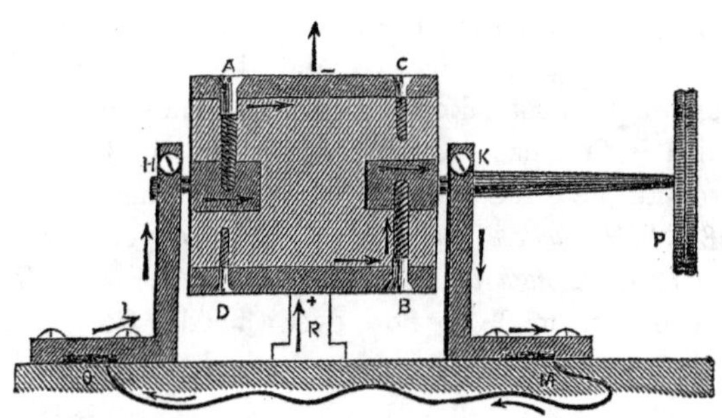

way into each end of the key. Each of these metallic pieces is connected by conductors with the copper strips running from the ends of the key. On opposite sides of the key are two pieces of copper (*a* and *a* in Figure 61), one of which is connected by a brass screw *A* to the metallic axis at one end of the key, and the other by *B* with the metallic axis at the other end. In one position of the key, the copper springs press against ebony, and the current is thus broken; in another position, the copper springs press against the pieces of copper on each side of the key, and the circuit is thus closed. The current then passes up the copper spring *R* connected with the positive pole of the battery, along the piece of copper *D B* on that side of the key, through the brass screw *B*, to the metallic axis *K* at one end of the key; thence, through the copper

Fig. 61.

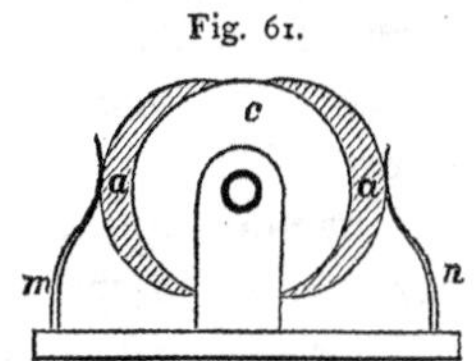

strips and wire *M*, to the other end of the key *O*, through the brass screw *A* at that end, to the piece of copper *A C* on the opposite side of the key, down the copper spring, and on to the negative end of the battery. Turn the key half-way round, and the other piece of copper on the key is brought against the spring connected with the positive pole of the battery, and the current must pass out from the key at the opposite end of the axis, and, through the remainder of the circuit, must take a course just the opposite of that in which it first passed. Hence, by turning the key half-way round, the direction of the current can be changed.

228. *A Magnetic Needle tends to place itself at right angles with a Wire through which a Current is flowing.* — If now the current be made to flow over a needle from its north end to its south end, the north pole of the needle will turn to the left hand of an observer who is facing that pole. If it be made to pass over the needle from its south pole to its north, its north pole will turn to the *right.* If it be made to flow *under* the needle, the north pole will turn in just the opposite direction; that is, to the right, when it flows from its north to its south pole, and to the left, when it flows from its south to its north pole. If a needle free to move in a vertical plane be held to the right of the wire, its north pole will point downward when the current flows from the north to the south pole of the needle, and upward when it flows in the opposite direction. If such a needle be held to the left of the wire, its north pole will point upward when the current flows from its north to its south pole, and downward when it flows in the opposite direction.

229. *Magnetic Rotations.* — We see, then, that the needle always tends to place itself at right angles to a wire through which a current is flowing. We see, moreover, that, if a needle which is free to move in both a horizontal and a vertical plane be held above a wire through which a cur-

rent is flowing from north to south, its north pole turns to the west; and that, if the needle be held on the west side of the wire, its north pole turns downward; and that, if it be held below the wire, its north pole turns to the east; while, if it be held to the east side of the wire, its north pole turns upward.

It would seem, then, that this pole would move round the wire, if it were free to do so.

The piece of apparatus represented in Figure 62 consists of a glass cup filled with mercury, an iron wire passing up through the bottom of the cup, with a bar magnet fastened to it by means of a string, and a copper wire dipping into the mercury above. As the mercury is a conductor of electricity, and offers no serious resistance to the movement of the magnet, the upper pole of the latter is free to move around the wire. When the iron and copper wires are connected with the opposite poles of the battery, the upper end of the magnet actually begins to rotate around the wire. If the upper end of the magnet is a north pole, and the copper wire is connected with the positive pole of the battery, the magnet rotates around the wire to the right; if the upper end of the magnet is a south pole, it rotates in just the opposite direction.

Fig. 62.

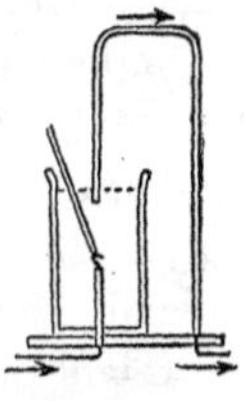

If the magnet be made stationary, and the wire movable, as shown in Figure 63, the wire will rotate around the pole of the magnet.

Fig. 63.

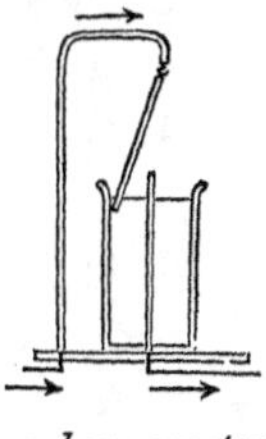

230. *The Rheoscope, or Galvanometer.* — Any instrument used for detecting or measuring a current is called a *rheoscope*, or a *galvanometer.* The first name means a *current-examiner;* the second, a *measurer of galvanism.* Current electricity is often called *galvanism*, from its discoverer, Galvani. Thus far, in our

examination of the existence and the strength of the electric current in the circuit, we have used a simple needle.

If this needle moves over a graduated arc, it will be found to move a greater number of degrees when the current passes entirely round it, than when it merely passes over it or under it. It is evident that this should be so. The current above and below the wire must, in this case, pass in opposite directions, and must, therefore, both tend to move the needle in the same direction. The effect is the same as if two currents of equal strength were passing over or under the wire, and both in the same direction. Every time, therefore, that the wire conducting the current is coiled round the needle, the effect of the current is multiplied. A current which is too weak to deflect the needle by simply passing over or under it, may be made to deflect the needle decidedly by coiling the conducting wire a great number of times round the needle.

231. *The Astatic Needle.* — When a single needle is deflected by the current, not only must the resistance caused by the weight of the needle be overcome, but also the *directive* action of the earth. It is this directive action of the earth which causes the needle to take its north and south position; and it offers a greater resistance to the deflection of a light needle than its weight does. This directive action can, however, be completely neutralized by the arrangement shown in Figure 64. Two needles of equal magnetic strength are fastened together, so that the north pole of one faces the south pole of the other. The earth will evidently pull each end of such a combination of needles both towards the north and towards the south with equal strength. It will therefore have no tendency to point north and south. Such a system of needles is called an *astatic*

Fig. 64.

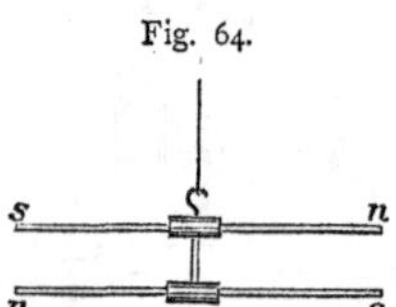

needle (from a Greek word meaning *unsteady*), that is, one *having no directive power.*

232. *The Astatic Galvanometer.* — If now copper wire be coiled a great many times around the lower needle, and an electric current be sent through the coil, it will act on both needles so as to turn both in the same way. For, suppose that the current passes over the lower needle from its north to its south pole, it will turn the north pole of that needle to the left. It will then pass under the upper needle from its south to its north pole, and will turn its south pole to the left of an observer facing that pole. If, then, an observer is facing the north pole of the under needle, the current between the needles will so act upon both as to turn the ends towards him to the left. As the current has only to overcome the resistance caused by the weight and the friction of the needles, they will be very sensitive to its action.

An instrument arranged as above described is called an *astatic galvanometer.*

When the needles are light, and delicately hung, such a rheoscope is exceedingly sensitive. It serves to detect and measure the feeblest current.

Fig. 65.

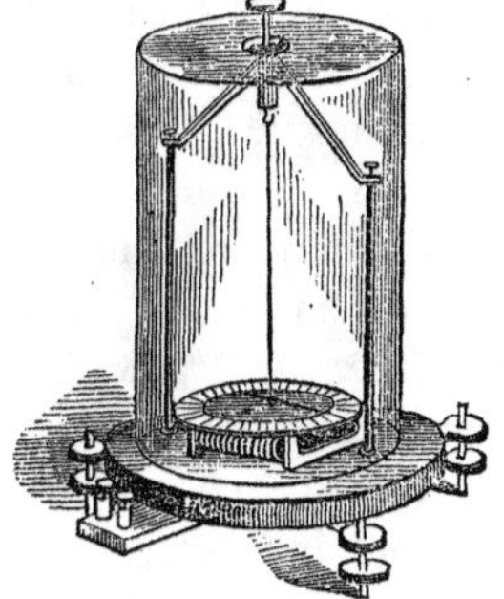

Figure 65 represents a galvanometer of this kind. The astatic needle is placed within a coil of fine copper wire carefully insulated with silk, and is suspended by a cocoon thread to a hook supported by a brass frame. It hangs freely without touching the coil, and the upper needle moves on a graduated circle. The whole is enclosed in a glass case, and rests on a stand supported by three levelling-screws.

233. *The Tangent Galvanometer.* — This instrument is shown in Figure 66. It consists of a thick strip of copper bent into the form of a circle from one to two feet in diameter, with a small needle moving on a graduated circle at its centre. It can be used for measuring the strongest currents; and, since the current in passing through the thick copper ring is resisted scarcely at all, it has the advantage of measuring the current without diminishing its strength.

Fig. 66.

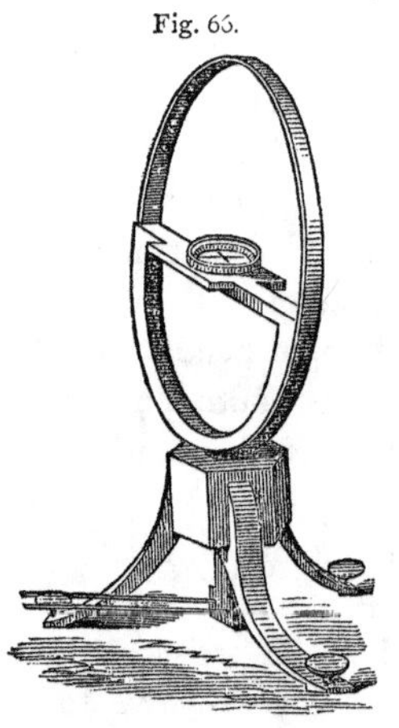

We have seen that the *astatic* galvanometer is used for measuring very feeble currents. For measuring currents of ordinary strength, a galvanometer of the same kind, but of less delicate construction, is used.

234. *The Needle Telegraph.* — Since the electric current passes with comparatively little resistance through thick conductors of almost any length, it is now much used in transmitting signals between distant stations. An instrument for sending such signals is called a *telegraph.* The word literally means *writing at a distance.*

Four things are essential in every kind of electric telegraph: (1.) a battery for generating the electricity; (2.) wires for conducting the electricity; (3.) an instrument for sending the message; and (4.) an instrument for receiving the message.

It has already been found that the opposite poles of the battery must be connected by a continuous conductor in order to obtain a current of electricity. It has been found that, if one end of the battery be connected with a copper plate sunk in the ground deep enough to be in contact with moist earth, and if a wire from the other end of the battery be connected with a copper plate similarly situated at a distant station, the current will pass just as well as when a wire passes from one end of the battery to the distant station, and then back to the other end of the battery. The earth in this case takes the place of one half of the conductor. In practice the circuit is always completed by the earth.

Suppose now, that at one station a rheotrope is introduced into the circuit, and at the other a galvanometer. Every time the circuit is closed by the rheotrope, the needle of the galvanometer is deflected; and every time the current is reversed, the motion of the needle is reversed. The operator at the rheotrope can thus readily make the needle at the distant station move any number of times either to the right or to the left. The galvanometer is placed in a vertical position, and the lower end of the needle is loaded, so that, when the current is broken, the needle quickly takes a vertical position. Out of the movements of the needle an alphabet can be readily arranged. Thus two movements to the left may represent *a*; three to the left, *b*; four to the left, *c*; one to the left and one to the right, *d*; and so on. When the alphabet has once been agreed upon, the operator by means of the rheotrope can easily send a message, and the operator at the distant station can, by noticing the movements of the needle, as easily read it.

Since this telegraph depends on the power of the current to turn a needle, it is called the *needle telegraph*. It was invented by Cooke and Wheatstone, and put into operation in England in 1837. It is not used in this country, but is still quite extensively used in England.

235. *Resistance to the Current.* — In the tangent galvanometer we have a ready means of measuring the strength of electric currents, and we find that the dimensions and the material of the substances included in the circuit exercise a

marked influence on the strength of the current. It is of the greatest importance, therefore, to ascertain the relative resistance of conductors of various forms and materials.

An instrument called the *rheostat* (that is, an instrument for making the *current steady*, or *of uniform strength*) is used for this purpose. It was invented by Wheatstone, and is constructed so as to introduce into or withdraw from the circuit a considerable amount of highly resisting wire, without stopping the current. It is shown in Figure 67. It consists of two cylinders, one of brass, the other of well-dried wood, turning on their axes by a crank. The wooden cylinder has a spiral groove cut into it, in which is placed a fine metallic wire; the brass cylinder is smooth. The end of the wire attached to the wooden cylinder is connected by means of a brass ring, with a binding-screw for the attachment of a battery wire. A metallic spring pressing against the brass cylinder is connected with the other binding screw. If now a current be sent through the wire, it will pass through all that portion of it which is wound at the time upon the wooden cylinder, but it will not pass through the portion wound upon the brass cylinder, but through the cylinder instead, since the latter is a better conductor than the fine wire. The wire wound upon the brass, then, is withdrawn from the circuit.

Fig. 67.

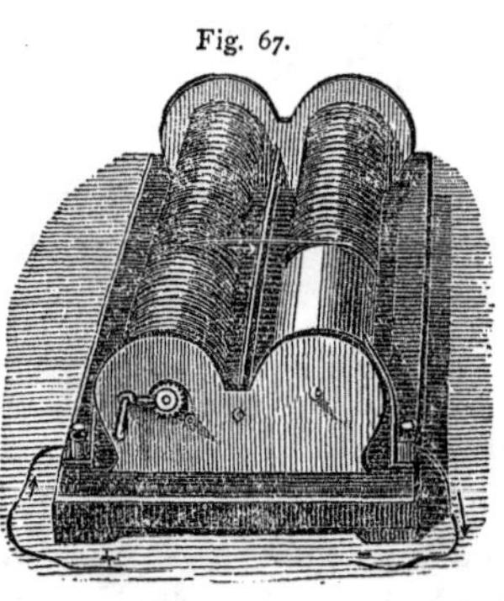

When the rheostat is to be used, all the wire is wound upon the wooden cylinder, and put into the circuit of a battery along with a galvanometer. If now the resistances of two wires are to be tested, the galvanometer is read before the first is put into the circuit. After it is introduced, the needle falls back in consequence of the increased resistance, and then as much of the rheostat wire is withdrawn from the circuit as will bring the needle back to its former place. The quantity thus withdrawn is shown by a scale, and is obviously equal in resistance to the wire introduced. The first wire is then removed, and the second wire is tested in the same way as the first. If 40 inches were withdrawn in the first case, and 60 inches in the second, the resistance offered to the current by the first wire is to that offered by the second as 40 is to 60; or, in other words, the former is two thirds of the latter.

By means of the rheostat it has been proved that *the resistances of wires of the same material and of uniform thickness are in the direct ratio of their lengths, and in the inverse ratio of the squares of their diameters.* Thus a wire of a certain length offers twice the resistance of its half, thrice that of its third, and so forth. Again, wires of the same metal, whose diameters are in the ratio of 1, 2, 3, etc., offer resistances which are to each other as 1, $\frac{1}{4}$, $\frac{1}{9}$, etc. Therefore, the longer the wire, the greater the resistance; the thicker the wire, the less the resistance. The same holds true of liquids, but not with the same exactness. The following, according to Becquerel, are the *specific resistances* of some of the more common substances, or the resistance which a wire of each, so to speak, of the same dimensions, offers at the temperature of 54° F.:—Copper, 1; silver, .9; gold, 1.4; zinc, 3.7; tin, 6.6; iron, 7.5; lead, 11; platinum, 11.3; mercury (at 57°), 50.7. For liquids the resistances are enormous compared with metals. With copper at 32° F. as 1, the following liquids stand thus: saturated solution of sulphate of copper at 48°, 16,885,520; ditto of common salt at 56°, 2,903,538; sulphate of zinc, 15,861,267; sulphuric acid, diluted to $\frac{1}{11}$, at 68°, 1,032,020; nitric acid at 55°, 976,000; distilled water at 59°, 6,754,208,000.

SUMMARY.

An instrument for breaking the current is called a *rheotome;* an instrument for changing the direction of the current, a *rheotrope.* (227.)

A magnetic needle tends to place itself at right angles to a wire through which a current is passing.

The direction in which a needle turns when brought near a wire through which a current is flowing, depends on the direction in which the current is passing, and upon whether the needle is held above, below, to the right, or to the left of the wire. (228.)

One end of a magnet free to move will rotate round a wire through which a current is passing. The direction of the rotation depends upon the direction of the current, and upon the nature of the pole which is free to move. If the magnet is fixed, and the wire through which the current flows is free to move, the latter will rotate round the former; the direction of the rotation depending upon the direction of the current, and upon the nature of the pole about which the wire rotates. (229.)

An instrument for indicating and measuring the current is called a *rheoscope* or *galvanometer.* (230.)

When a simple needle is deflected by a current, its tendency to remain at rest, and the magnetic attraction of the earth, must be overcome. The latter resistance is neutralized in the *astatic* needle, which is therefore more sensitive to the action of the current than a simple needle. (231.)

When a current passes entirely round the needle, it deflects the latter as much as two currents of the same strength and direction flowing over the needle. Hence, the effect of the current upon a needle is multiplied by coiling the wire round the needle. (230.)

The *astatic galvanometer* consists of an astatic needle with a coil of wire round the lower needle of the system.

The *tangent galvanometer* consists of a small magnetic needle in the centre of a circle of thick copper wire.

A very delicate astatic galvanometer is suitable for measuring very weak currents; a tangent galvanometer for measuring very strong currents. For currents of intermediate strength an astatic galvanometer less delicately made is suitable. (232, 233.)

A *telegraph* is an instrument for writing at a distance. Four parts are essential to every telegraph: (1.) a battery for generating, and (2.) a wire for conducting the current; (3.) a sending instrument, and (4.) a receiving instrument.

The *needle telegraph* depends upon the power of the current to turn a needle. The sending instrument is a rheotrope, and the receiving instrument a galvanometer. The letters are indicated by the movements of the needle. (234.)

The relative resistance offered to the current by different conductors is measured by means of the *rheostat.*

The resistances of wires of the same material and thickness are directly as their lengths, and inversely as the squares of their diameters. (235.)

ELECTRO-MAGNETISM.

236. *The Current can make Iron magnetic.* — If a part of the wire of the circuit be wound into a coil, and a piece of soft iron be placed inside this coil, it becomes strongly magnetic while the current is passing; as may be shown by bringing bits of iron near the ends of the iron inside the coil. Such a magnet is called an *electro-magnet.* The coil

is called a *helix*. If the coil is a left-hand coil (see Figure 68), the end at which the current enters the coil will be

Fig. 68.

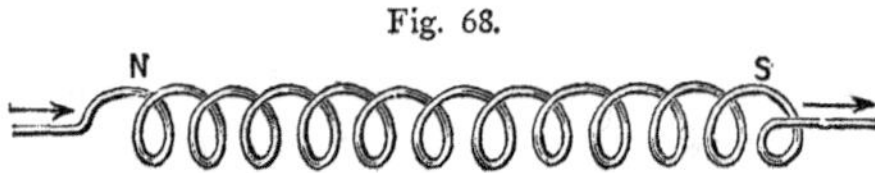

found by means of the magnetic needle to be the north pole. So that, by reversing the current, the poles of the electro-magnet will be reversed. If the coil is a right-hand

Fig. 69.

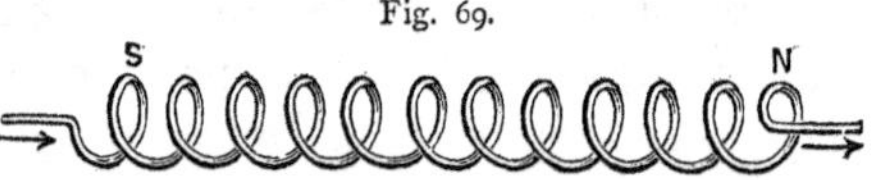

one (see Figure 69), the end at which the current enters is found to be the south pole.

When the current is broken, the soft iron instantly loses its magnetism, as is shown by the falling of the bits of iron which before clung to it. If a steel rod is used, instead of a soft iron one, it retains its magnetism after the current is broken. If the wire is wound around the iron in several layers, the strength of the magnet is greatly increased.

Steel magnets are now usually made by means of the electric current. A short, thick coil of copper wire and a powerful battery are used for this purpose. A steel bar is inserted into the coil, and moved backwards and forwards, and when the middle part is a second time in the coil, the circuit is opened, and the bar may then be withdrawn, perfectly magnetized.

The strongest electro-magnets are of the horseshoe form. They far exceed ordinary magnets in power. Small electro-magnets have been made by Joule in England, which support 3500 times their own weight; and a large one was constructed by Professor Henry, of the Smithsonian Institution, which supports a weight of 2500 pounds. These magnets are much stronger when provided with a *keeper*, or *armature*, that is, a piece of soft iron which connects the two poles, as shown in Figure 70. Electro-magnets are sometimes mounted in frames (see Figure 71), which is a convenient form for showing their strength.

Fig. 70.

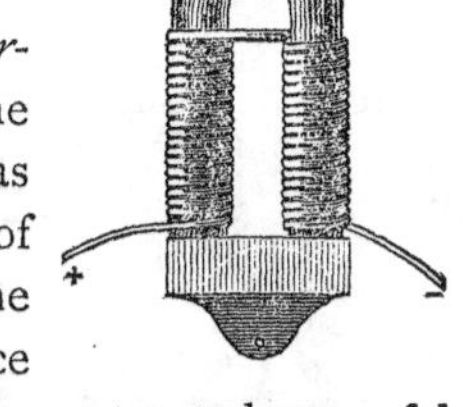

237. *The Wire through which a Current is passing is a Magnet.* — If the current be sent through a coil such as is shown in Figure 72, and the end of a rod of soft iron be brought near the opening in the centre, it is at once drawn into the coil. Coils have been constructed powerful enough to draw up and sustain a weight of 600 pounds.

Fig. 71.

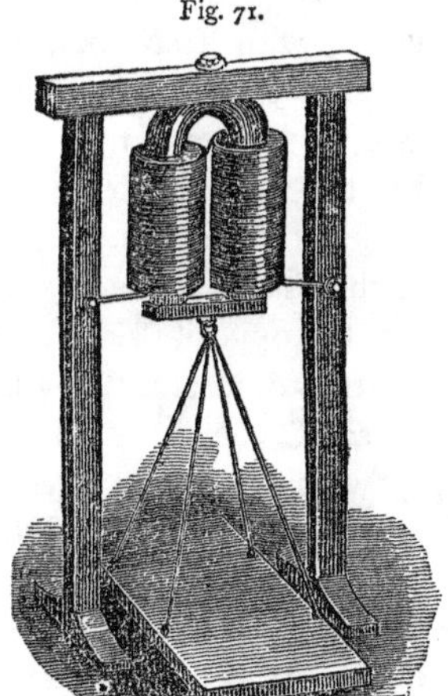

Fig. 72.

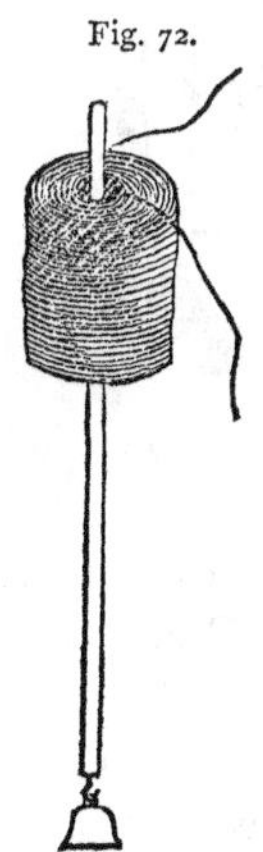

The electric current, then, is not only able to develop magnetism in soft iron, but the coil itself, through which the current is passing, is magnetic. If the wire which joins the poles of a battery is brought in contact with fine iron-filings, they adhere to the wire; showing that any wire through which the current is flowing is magnetic.

238. *Electricity as a Source of Mechanical Power.* — A great variety of electro-magnetic machines have been constructed with a view to applying electric force to the work now done by steam. They all depend on the property of an electro-magnet instantly to acquire or to lose its magnetism on the passage or the interruption of the current; or to reverse its poles when the direction of the current is changed.

Page's rotating machine, shown in Figure 73, illustrates one method of applying the electric force to do mechanical work. It consists of a horseshoe magnet, in the axis of which is an upright shaft. On this shaft, and at right angles with it, is fixed a piece of soft iron, with its ends facing the poles of the magnet. The soft iron is surrounded with a coil of copper wire, so that it is an electro-magnet. The ends of the wires of the coil are fastened to two metallic strips, which are so formed that each nearly half surrounds the shaft. These strips are fastened to the opposite sides of the shaft, but separated from it and from each other by a non-conducting substance. The current comes to the coil through two metallic springs, which press against the strips. When the shaft has performed half a revolution, the springs pass from one strip to the other, so that the current is made to enter the coil at the end of the wire opposite to that which it entered at first; and so the poles of the electro-magnet are reversed.

Fig. 73.

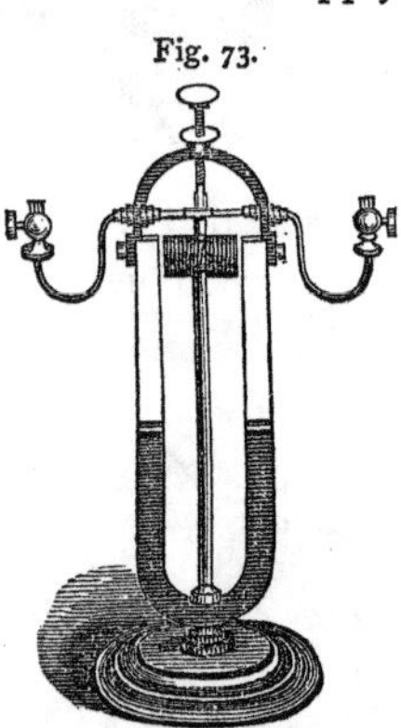

The machine is so arranged that, at starting, the poles of the two magnets facing each other are of the same kind. They therefore repel each other, and when the shaft is once started, they send it around a quarter of the way; then unlike poles begin to approach each other, and their attraction causes the shaft to complete half of a rotation. The current then changes its direction through the coil, the poles of the electro-magnet are reversed, and like poles again face each other and are repelled. The rotation is kept up by the self-acting rheotrope. Such a shaft may be made to rotate 2000 times a minute, causing 4000 reversals of polarity in that short time. The horseshoe magnet may, of course, be an electro-magnet; and then engines of considerable power may be constructed.

One of the best electro-magnetic engines is that devised and constructed by Froment. It is shown in Figure 74.

Fig. 74.

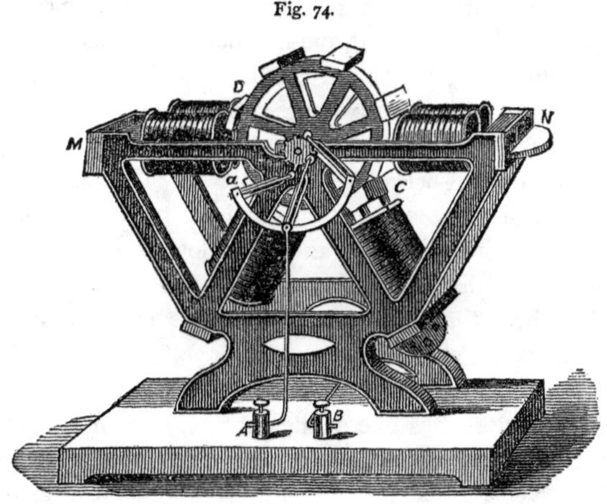

It consists of six electro-magnets fastened to an iron frame. Only four of these magnets appear in the figure: there are two above, corresponding to the two below. The poles of these magnets face a wheel, upon whose rim are eight pieces of soft iron, placed at equal distances. As the magnets are also placed at equal intervals, only two of the pieces of soft iron can be opposite the poles of the magnets at once. As the wheel stands in the figure, one of the pieces of iron is just above the magnet seen on the right, and another just below the magnet on the left. If now a current passes through the coils of these magnets, they become active, and each draws the piece of iron nearest it towards itself; and the wheel is thus made to turn till the pieces of iron are brought directly opposite the poles of the magnets, thus carrying the next pieces of iron very near the next magnet. If now the current is stopped in the first pair of coils, and sent through the next, the next pieces of iron will be drawn opposite the poles of these magnets; thus carrying the wheel onward, and bringing the next pieces of iron near the next magnets. The current is then stopped in the magnets through which it was last sent, and is sent through the coils of the next, and so on. In this way the wheel is made to rotate rapidly.

The current is sent through the coils in succession, by means of the cog-wheel seen at the end of the axis of the wheel. One pole of the battery is connected with this wheel, and the other with the arc below, which carries the three springs that are seen to press against the cog-wheel. As the wheel turns, the teeth strike the springs one after another, and thus close the circuit. The coils of the electro-magnets are connected in such a way, that every time a tooth touches one of the springs, the current passes through the coils of two magnets directly opposite each other. As soon as the pieces of soft iron are drawn opposite the poles of these magnets, the tooth leaves this spring, and another tooth touches the spring which closes the circuit in which the coils of the next pair of magnets are included, and so on.

Electro-magnetic engines have never yet been constructed of above eight or ten horse-power, though there is apparently nothing to limit them to this low power. The great obstacle to the success of these engines is the expense of generating the electricity to run them. It costs some forty or fifty times as much to generate electric force as to generate the same amount of steam force. Yet, for certain kinds of work, where rapid motion and comparatively little force are required, electric engines have been found to answer better than small steam-engines.

239. *Foucault's Self-acting Rheotome.* — This instrument, shown in Figure 75, illustrates one of the many applications

Fig. 75.

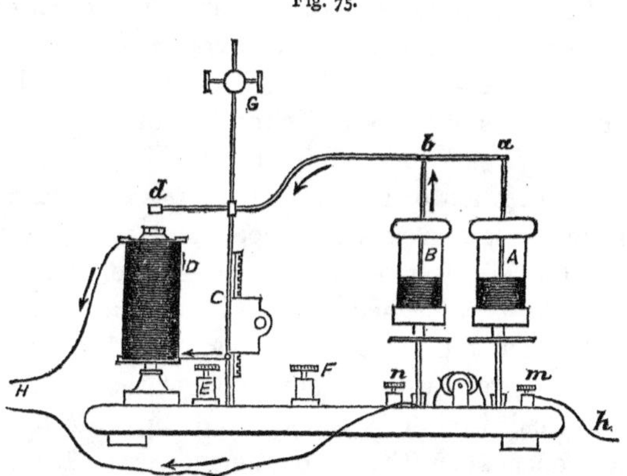

of the electric force to doing mechanical work. It consists of a beam, *a d*, supported by a standard, *C G*, which acts as a spring. At one end of the beam there is a keeper of soft iron; at the other end, two iron rods, which plunge into cups *A*, *B*, partially filled with mercury. Under the iron keeper is an electro-magnet, *D*. One end of the wire of the helix of this magnet connects with one pole of a Bunsen's cell. The other pole of this cell is connected with the mercury cup, *B*. The other end of the wire of the helix is connected with the beam by means of the standard; so that the circuit of the Bunsen's cell is closed when the iron rod dips into the mercury, and is open when it is out of the mercury. It is best to cover the mercury with alcohol, which is a non-conductor.

When the rheotome is to be worked, the iron rod is so adjusted that its end is just above the surface of the mercury. That end of the beam is then depressed by the hand so as to bring the rod into the mercury. This closes the circuit, and renders the electro-magnet active, and the keeper at the end of the beam is drawn down upon it. This carries the other end of the beam up and the rod out of the mercury, opens the circuit, and renders the electro-magnet inactive. The elasticity of the standard throws this end of the beam back and lowers the rod into the mercury, closing the circuit again. Everything is now as at the first, and the same succession of movements is repeated indefinitely.

This instrument is made to open and close a second circuit in the following manner. One pole of the battery of this circuit is connected with the beam, and so with the iron rod, which dips into the second cup of mercury, *A*, which is connected with the other pole of the battery; so that this circuit is closed when the rod dips into the mercury, and open when it is out of the mercury. But if the point of the rod is so adjusted as to be just above the surface of the mercury, it is drawn out of it every time that the keeper is drawn down to the electro-magnet, and is plunged into it every time that the keeper is thrown back by the spring.

Fig. 76.

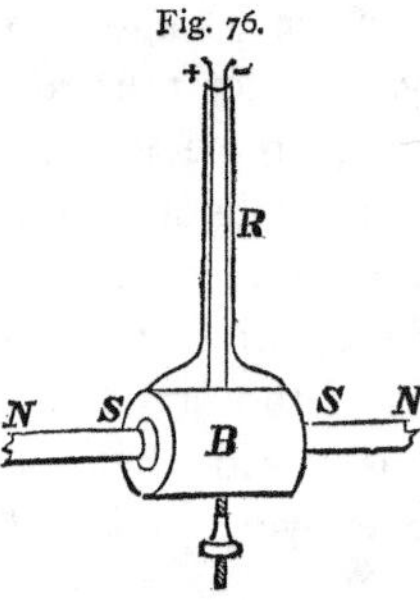

240. *Electric Clocks.* — The electric force has also been used to regulate the movements of clocks. The clocks thus regulated are called *copying clocks*. They are of the usual construction, except that the pendulum balls are hollow coils of copper wire, so that they become magnetic when a current is sent through them. In Figure 76, *R* represents a part of the

rod, and *B* the ball, of such a pendulum. Permanent magnets, *NS* and *SN*, are fastened against the sides of the clock-case opposite the ends of the coil *B*, with like poles towards the coil. The hollow of the coil, as it swings, can pass a little way up the length of each magnet. If the south poles of the magnets are turned towards the coil, as in the figure, and a current is sent through the wire, one end of the coil becomes a north pole, which is attracted by the magnet near it, and the other end a south pole, which is repelled by the magnet near it. This attraction and repulsion both tend to send the coil in one direction. If, now, at the instant that *B* is drawn to one side, the direction of the current is changed, the poles of the coil are reversed, and it is carried to the other side. The pendulum thus vibrates every time the current is reversed. This is done by means of a standard or regulating clock. Every time the pendulum of this clock vibrates, the direction of the current is reversed; so that the pendulums of all the copying clocks vibrate exactly at the same rate as the pendulum of the regulating clock. In this way, by means of one accurate clock, any number of copying clocks, of the most ordinary construction, can be made to keep accurate time.

Fig. 77.

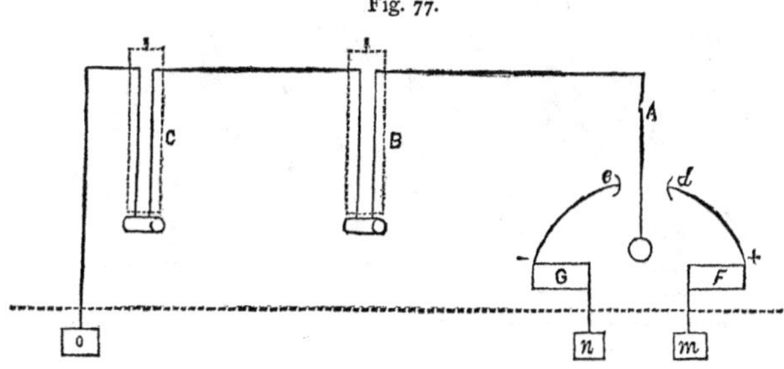

Figure 77 shows one of the ways in which the pendulum, *A*, of the regulating clock can change the direction of the current. The spring *e* is connected with the negative pole of the battery *G*, and the spring *d* with the positive pole of the battery *F*. The other poles of these batteries are connected with the plates *m* and *n*, buried in the earth. *B* and *C* are the pendulums of the copying clocks. When the regulating pendulum touches the spring *d*, the current flows through the wire from *A* to *B* and *C*; when it touches the spring *e*, the current flows first through the earth from *n* to *o*, and then through the wire from *C* to *A*. The permanent magnets connected with the pendulums *B* and *C* are not represented in the diagram.

241. *Morse's Electro-magnetic Telegraph.* — The telegraph now most extensively used, both in this country and in Europe, is the one constructed by Morse in 1841. This telegraph depends on the power of the current to develop magnetism in soft iron, and is hence called the *electro-magnetic* telegraph. Its importance demands a somewhat detailed description.

Fig. 78.

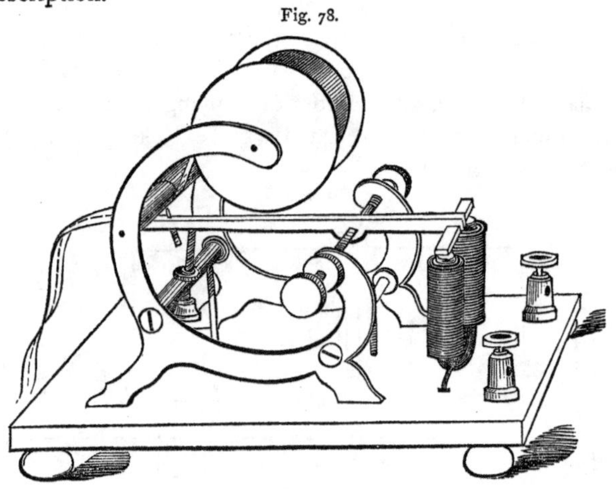

242. *The Receiving Instrument.* — The essential parts of the receiving instrument are shown in Figure 78. One of

the screw-cups at the right is connected with the line wire from the distant station, and the other with the earth. The current traverses the coils of the electro-magnet, and draws down the keeper and the arm of the lever to which it is attached. The other end of the lever is raised, pressing a steel point, or style, against a strip of paper, which is unrolled from the bobbin above, and moved steadily along by clock-work not represented in the figure. When the current from the distant station is broken, the shorter arm of the lever is released by the electro-magnet, the longer arm falls back by its weight, and the style ceases to press against the paper. The kind of mark made upon the paper depends upon the time the style remains elevated. If it is raised for a moment only, a dot is made; if for a longer time, a dash. The alphabet used is made up by the combination of dots and dashes. The following is the usual Morse alphabet:—

A . —	J . — — —	S . . .
B — . . .	K — . —	T —
C — . — .	L . — . .	U . . —
D — . .	M — —	V . . . —
E .	N — .	W . — —
F . . — .	O — — —	X — . . —
G — — .	P . — — .	Y — . — —
H	Q — — . —	Z — — . .
I . .	R . — .	

The alphabet is so arranged that the letters occurring most frequently are most easily signalled; thus, E is one dot; T, one dash. An expert operator can transmit from thirty to forty words a minute on a land line of from 200 to 300 miles.

A clerk accustomed to a Morse telegraph seldom looks at the paper in transcribing. The mere clicking of the lever becomes a language perfectly intelligible to him. He need therefore look at the record only when he may have heard indistinctly.

243. *The Sending Instrument.* — Let us now transfer our attention to the distant station, to see how the current is transmitted from it. The sending instrument, or *transmitting key*, is shown in Figure 79. A brass lever, *l l*, moves on the axis *A*. It has two projections of platinum, *m* and *n*, on its lower side. These strike

Fig. 79.

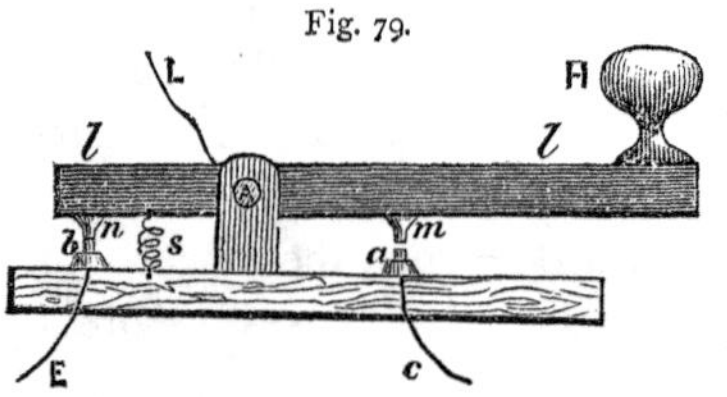

against pieces of platinum *b* and *a*, the first of which is connected with the earth-wire *E;* the second, by the wire *c*, with one of the poles of the sending battery. When the lever is left to itself, *n* and *b* are in contact under the force of the spring *S*. When the hand presses on the ebony handle *H*, contact is broken at *n* and *b*, and established at *m* and *a*. Besides the wires *E* and *c* already mentioned, the line-wire *L*, from the distant station, is connected with the lever, through its axis, *A*. When the key is in the receiving position (that shown in the figure), the current from the sending station takes the route *L*, *A*, *l*, *n*, *b*, *E*, the recording instrument, then to earth. When *H* is pressed down, the key is in the sending position, and transmits the battery current by *c*, *a*, *m*, *A*, *L*, to the distant station.

244. *How two Stations are connected.* — Figure 80 shows how two stations are connected on Morse's system. *B* and *B'* are the batteries at the stations *S* and *S'* ; *k k'* are the transmitting keys ; *n n'*, the receiving instruments ; *g g'*, the galvanometers ; *L L*, the line-wire insulated on posts ; *P P'*, the earth-plates. When the key *k*, at the station *S*, which is here represented as the sending station, is depressed, the current from the battery *B* takes the following course.

Fig. 80.

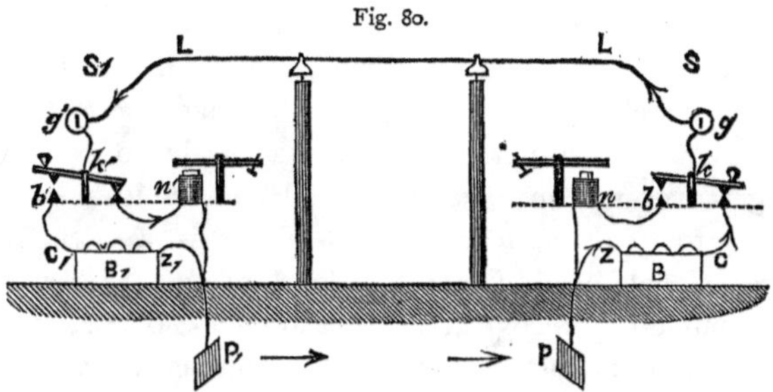

From the positive pole C of the battery B, through k, to the galvanometer g; then through the line L L to the receiving station S'; through the galvanometer g', the key k', the coils of the receiving instrument n'; thence to the earth at the plate P', by the ground at P to the sending station; and finally to the negative pole Z of the battery B. At station S, b and n are out of circuit; and at S', b' and B' are out of circuit; n is thrown out of circuit, because its coil offers a resistance equal to several miles of the line-wire. If it were in circuit, both registers could print at the same time; but that is not necessary, one record at the receiving station being enough. The galvanometer g serves to reveal the currents put in circuit, and thus to show when a message has been transmitted. It also shows the presence of earth-currents on the line. If k were left to itself, and k' depressed, the station S' would then be the sending, and S the receiving station; and the connections would be exactly as shown in the figure, only at opposite stations.

If the clerk at S wishes to telegraph to S', he depresses the key k several times, so as to send a series of dots and dashes giving the name of the station. The attention of S' is first arrested by the clicking of the armature of his receiving instrument. He sends back word to S that he is ready, and the printing then begins. When both keys, k and k', are depressed, the whole circuit is broken; so that

9*

when sender and receiver have their hands on their respective keys no message can be sent. One might fancy that confusion would arise from cross-messages; but operators soon get over this inconvenience, and communicate back and forward with perfect facility. There is a code of working signals to indicate the kind of message, etc., besides numerous recognized contractions. To arrest the attention of attendants, the current is sometimes made to ring an alarm-bell.

245. *The Line.* — Telegraphic stations must be united by one insulated wire, carried either over land or under the sea. The insulation of land lines is insured by attaching the wires to insulators fixed on posts some twenty feet high. The posts are placed at distances of from eighty to one hundred and fifty yards, according to the number of wires they have to carry.

Insulators are of a great variety of shapes. A very good form is shown in Figure 81. A bolt of iron, *I*, fixed to the cross-beam, *w*, of the telegraph pole, is cemented into the ebonite cup or bell *e e*, which, in turn, is cemented into the porcelain bell *p p*. The line-wire is kept fixed by binding-wire into the groove *a*. The electricity has thus to travel over the outside and inside of both cups before it reaches the iron bolt, which is also coated with ebonite. The leakage on a long line, notwithstanding the best insulation, is considerable. The loss at each post is insignificant; but when hundreds or thousands are taken into account, it becomes decided, so that merely a fraction of the original current reaches the earth at the distant station. In rainy, and more especially in misty weather, the insulation suffers much.

Fig. 81.

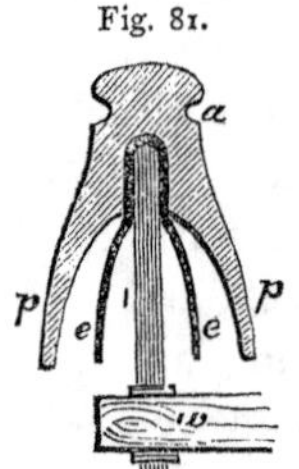

246. *Submarine Lines.* — A submarine line is made by a cable. The core of the cable consists of one wire, or a

strand of several wires, of copper, as pure as can be got in the market. One solid wire is preferable to a strand of the same diameter in point of conducting power, but a strand is surer; for, when one wire is broken at a point, the others still remain to conduct the current. When the one solid wire is broken, which may happen without its being visible outside, the cable becomes useless. The strand of wire is generally covered with a compound of gutta-percha and resinous substances, which fills the interstices between the wires. It is then included in one or more coatings of gutta-percha, then in a layer of tarred yarn, and finally in a sheathing of iron wire, laid on spirally, to give the cable sufficient strength to withstand the strain of paying out, or that to which it may be subjected from the inequalities of the ocean bed. Figure 82 shows the construction (full size) of the Malta and Alexandria cable, 1330 nautical miles long, and one of the best in operation. *O* is a strand of seven copper wires, laid in a compound of gutta-percha and resins; *C*, three layers of gutta-percha, with the same compound between them; *H*, tarred yarn; and *I*, the eighteen wires constituting the sheathing. The diameter out in the sea is .85 of an inch. Near the shore, where it is more exposed to injury, the sheathing is made much stronger.

Fig. 82.

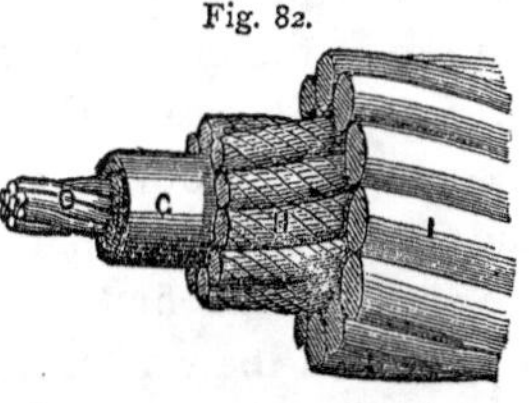

Considerable dispute has arisen as to the best material for insulating marine cables. India-rubber and gutta-percha are the two rival substances. It may be said in favor of gutta-percha, that not one yard of it, when laid, has decayed; and that, under ocean pressures, as proved by the Atlantic cable of 1865, its insulating power decidedly improves. In favor of India-rubber it is urged that cables alike in other respects will, when coated with it instead of gutta-percha, transmit twice as many words per minute. On the other hand, India-rubber has not proved so durable, some specimens of it having become treacly after immersion for some time in the sea.

247. *The Earth.* — One wire is quite sufficient to connect two telegraph stations, if its terminations be formed by two large plates sunk in the earth. The plates are generally of copper, and should have a surface of not less than twenty square feet; and they must be buried so deep that the earth about them never gets dry. The gas and water pipes in a town make an excellent *earth*, or earth-connection. When the *earths* are good, the current passes through the earth between the two stations, no matter what may be the nature of the region it has to pass, — plain or mountain, sea or land. The resistance of the earth to the current, compared with that of a long line, is next to nothing. The earth serves the purpose, not only of a second wire, but of one so thick that its resistance may be left out of account. In conducting power, for equal dimensions, the earth stands much inferior to the wire; but then its thickness, so to speak, is indefinitely greater, and hence its conducting power, on the whole, is superior.

248. *The Relay.* — It is only on short circuits, generally of less than fifty miles, that the receiving instrument is worked directly by the line current. On long circuits, direct working could only be accomplished by an enormous sending battery. The loss by leakage on the way is very considerable; so that a current strong at starting becomes very weak before it reaches the station to which it is sent. Besides, the leakage is the greater, the greater the number of cells employed, or the greater the tension of the battery. It is found a much better arrangement to work the receiving instrument by a local current, and to include a very delicate instrument in the line circuit, which has only to make or break the local circuit. Such an instru-

ment is called a *relay*, and is shown in Figure 83. The electro-magnet, *E*, of the relay is included in the line circuit, instead of the electro-magnet of the receiving instrument. The coil is long, and of very fine wire; and a very faint current is sufficient to develop magnetism in the core. The keeper, *A*, of the relay is attached to a lever, *ee'*, turning on the axis *a*. When a current is sent through the coil, the lever is drawn down, and the end *e'* rests on the screw *S*. When there is no current, the elasticity of the spring *s* brings it back against the screw *S'*. The pillars *N* and *P* are connected with the poles of the local battery. The metal spring *s* places the lever *ee'* in connection with *P*. The screw *S* and the end *e'* of the lever, then, are virtually the poles of the battery. When these are in contact, the local current flows, and it stops when *e'* is brought back against the insulated screw *S'*. The receiving instrument is included in the local circuit. When a current comes from the sending station, the keeper *A* is attracted, *e'* falls on *S*, the local circuit is closed, and the receiving instrument begins to print. When the current ceases, *e'* returns to *S'*, and the style of the receiving instrument is withdrawn from the paper. The effect is thus the same as if the line current printed, and not the local current. By this means, a current too weak to work the receiving instrument can complete the local circuit and print legibly.

Fig. 83.

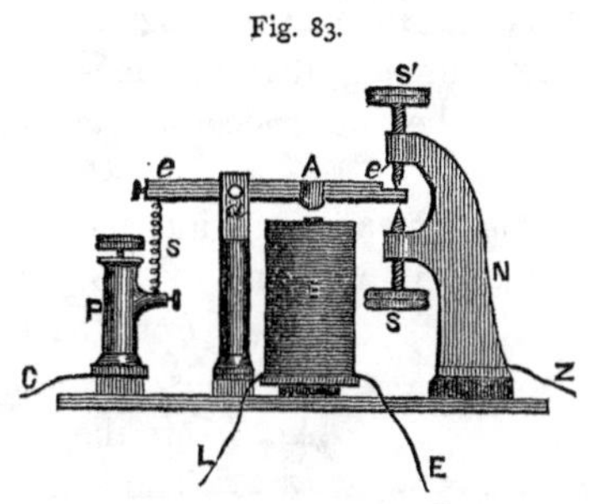

249. *How several Stations are connected in one Circuit.* — This is effected in three ways: (1.) by an *open circuit;* (2.) by a *closed circuit;* and (3.) by *translation.* In all of these, each station may telegraph simultaneously to all the sta-

tions in the circuit; and, if the message concerns them all, a record may be printed at each station. When a station wishes to telegraph to another, it keeps signalling the name till the station in question signals back that he is ready. The others, finding that the message does not concern them, leave the two concerned in possession of the circuit.

The arrangement of an intermediate station in an *open circuit* is shown in Figure 84. L_1 and L_2 are the wires from the terminal stations; R is the relay; the other parts are the same as in Figure 80. The station is represented as receiving. The line current passes through the key and the relay, and goes on to L_2. The relay sets the local battery and the receiving instrument in operation. The line current is brought into the station and led out without being affected. Electrically, it is the same as if it had gone on in the air direct from L_1 to L_2. When the station is sending, the key is depressed. The current goes from C into the line L_1, is *earthed* at the one terminal station, leaves the earth at the other, and returns to Z by L_2. The battery here has no *earth*, as at the terminal stations, the arrangement of which is as in Figure 80. An *earth*, however, is generally put at each station, so that it may be worked as a terminal station, if required. R, at sending, is out of circuit. According to this plan every station must have a battery as strong as at the terminal stations.

Fig. 84.

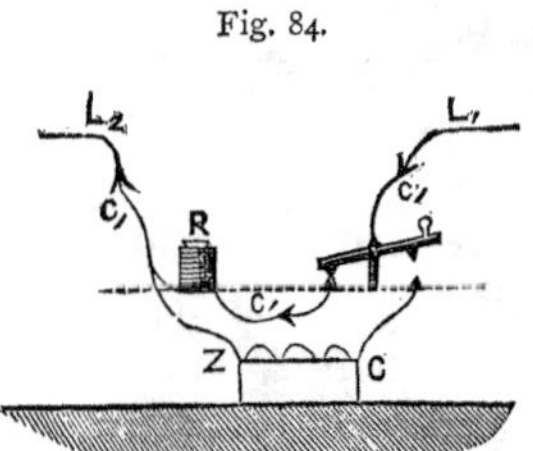

In the *closed circuit*, no battery is needed at the intermediate station. If the battery and its connections be removed, Figure 84 gives the arrangement in a closed circuit. The battery may be placed only at one terminal station, or it may be divided, and a half placed at each end; both, however, being so arranged as to act with, not against, each other. The circuit is closed when no one operates, so that a current is constantly flowing. The keys, breaking the connection, stop it for the time. The relays act negatively, making the receiving instrument print when there is no line current, and rest when it flows. If S', in the relay (Figure 83), were uninsulated, and S insulated, it would act in a closed circuit. The advantage of the closed circuit is, that the batteries, which require considerable attention, are confined to the terminal stations, where they can be best cared for. Besides, little or no adjustment is required for the relays. All the relays are in circuit at once.

Open and closed circuits are used on lines where a number of smaller towns are joined together, the business of all of them being no more than sufficient to keep the line working. They are for short distances, seldom more than 200 or 300 miles.

When two stations, say 500 miles apart, are to be connected by telegraph, it is seldom done by a direct line, but by transmitting to a half-way station, and thence to the terminal one. This method is called *translation*. It is effected by making the lever of the receiving instrument act as a relay in transmitting the message to the next station. Figure 85 shows how this may be done. The current C_1, from the sending station, enters the coil M of the receiving instrument, and goes thence to earth at P, and returns as shown by the arrow C_1. The instrument may print or not, according to the message, but its doing so or not interferes in no way with the translation. The copper pole C of the battery is connected with the lever $l\,l'$

Fig. 85.

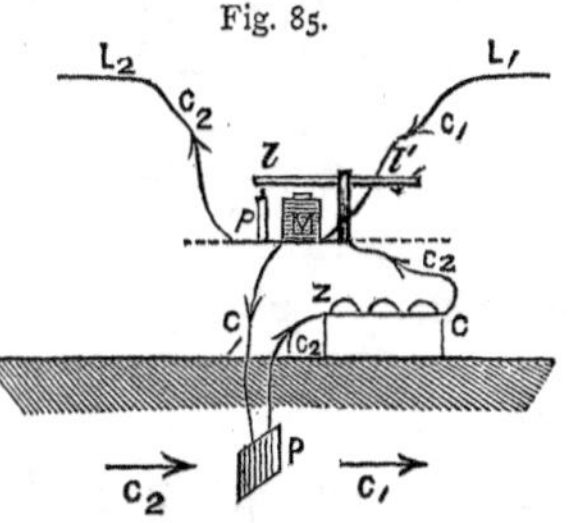

of the receiving instrument, and the zinc pole with the earth. When the lever is drawn down by the current C_1, it strikes against the point at the top of the pillar p, which checks its motion. This pillar is joined to the line L_2, running to the farther station, and when the lever falls, a second circuit — namely, that of the battery — is closed, in which C, the lever, the pillar, L_2, the farther station, the earth, P, and Z are all included. Thus as ll' prints at the intermediate station, it at the same time sends a new printing current to the next. When it ceases to print, the printing ceases at the distant station.

At the shore ends of submarine cables there is always a translating apparatus. This allows the cable to be worked by a battery suited to it, without loss of time in making it a special circuit.

250. *The Combination Printing Telegraph.* — The Morse recording instrument, as we have seen, writes by means of dots and dashes. An instrument has been invented by Mr. Royal E. House, a native of Vermont, which records the message in plain Roman letters. This telegraph is known as House's Printing Telegraph, and was patented in 1848. It has been found to work well, and has been used on many lines.

In 1855, Mr. David E. Hughes of Kentucky, after ten years of persevering labor, produced a printing telegraph on a new principle, simpler in construction and capable of working upon long circuits. In this telegraph each electric impulse sent over the wire prints a letter, while the House instrument requires on an average seven impulses for each letter, and the Morse an average of three and a half impulses.

Both of these printing telegraphs have now been in great measure superseded by the Combination Telegraph devised by Mr. Phelps of Troy, N. Y. This instrument, as its name indicates, is a combination of the principles of the two preceding telegraphs, with certain improvements, originated by Mr. Phelps. It is very simple in principle and rapid in action, and is likely soon to supersede even the Morse telegraph.

The sending instrument is a rheotome of peculiar construction, somewhat resembling a piano in outward appearance. On the key-board are twenty-eight keys, upon which are printed the twenty-six letters of the alphabet, a dot, and a dash. Near the key-board is a brass cylinder. On each key there is a peg, and in the cylinder there are twenty-eight cavities, one for the peg of each key, so arranged that each peg can enter its own cavity and no other. The cavities are arranged spirally around the cylinder, so that each cavity is $\frac{1}{28}$ of the circumference of the cylinder behind the preceding one. The cavity is so formed that the peg of the key, on entering it, is carried a little to the left, and thus completes the circuit. If all the keys were depressed at once, the circuit would be closed twenty-eight times at equal intervals during one rotation of the cylinder. If a key at the beginning of the alphabet, and another at the middle, were depressed and kept depressed during one rotation, the circuit would be closed twice, and the interval between the closings would be that of half a rotation. By means of this instrument the circuit can be closed with great precision at fixed intervals.

At the receiving station there is a small disc, called the *type-wheel*, on the edge of which are types of the twenty-six letters and the dot and dash, arranged at equal intervals. By the side of the type-wheel is a little press just large enough to take off one letter from the type-wheel. This press is forced against the type-wheel by machinery. When this machinery is at rest, the press is thrown back from the type-wheel by a spring. A strip of paper is carried along between the type-wheel and the press at such a rate that it advances the width of a letter every time the press is pushed against the wheel.

The type-wheel and press are moved by clock-work, as is also the printing cylinder at the sending station. The clock-work at each station is regulated by the vibrations of springs which are capable of vibrating the same number of times a minute. By this means the printing cylinder and the type-wheel are made to move at exactly the same rate. The two instruments are so set that the letter *A* on the type-wheel is opposite the press at the same instant that the cavity in the printing-cylinder corresponding to that letter is under the peg of its key. If the key were depressed at this instant, the circuit would be closed, an electric impulse would be transmitted, and, in passing through the coil of the electro-magnet at the receiving station, would render it active and cause it to draw down its keeper. To this keeper is attached a *detent*, or catch, which arrests the motion of the machinery that works the press. When the keeper is drawn down by the magnet, this detent is withdrawn, the press is driven forward, and the letter opposite is printed on the paper. When the circuit is again opened, the keeper is thrown back by means of a spring, the detent replaced, and the press removed from the type-wheel by the spring arranged for that purpose. Since both the type-wheel and the cylinder rotate at the same rate, it is evident that, when the peg of any key enters its cavity, the letter of that key will be printed at the receiving station. Every time the circuit is closed — that is, every time a key is depressed — the detent is withdrawn from the wheel that moves the press, and a letter is printed on the paper. The type-wheel rotates at the rate of about one hundred and twenty times a minute. The ordinary speed of this instrument is two thousand words an hour, which is about twice as fast as the Morse can work.

251. *The Telegraphic Fire-Alarm.* — The electric telegraph is now extensively used for indicating the locality of fires in cities. In various parts of the city are small iron boxes called *signal-boxes.* In Boston there are some seventy-five of these boxes. They are all numbered, and connected with a central station by means of wires. By turning a crank which is found inside the signal-box, the circuit is opened and closed in such a way as to telegraph to the central station the number of the box. When, therefore, a fire occurs in the neighborhood of any box, the box is opened, the crank turned, and the number of the box telegraphed to the central station. This station is also connected by wire circuits with several bells in different parts of the city, and the operator, by means of the electric force, rings on these bells the number of the box near which the fire is, so that the firemen know at once almost the exact locality to which they must go.

The hammers which strike the bells are worked by weights, the machinery being similar to that of the striking apparatus in an ordinary turret clock. The train of wheels is kept from moving by a detent, or catch. When the current passes, it develops magnetism in an electro-magnet, which attracts a keeper in front of it. This keeper supports a small lever poised nearly vertically, and weighted with a little ball near its upper end. This lever is tripped by the withdrawal of the keeper, and in falling acquires sufficient force to strike up the detent. The machinery is thus set in motion, and the hammer strikes the bell. A single blow of the hammer follows each electrical impulse, and the motion of the machinery raises the lever again to its place, and poises it on the keeper ready to be tripped for another blow. If the number of the box is ten or less, it is indicated by a corresponding number of strokes on the bell. If above ten, the digits of the number are indicated by striking the numbers corresponding to them with a short pause between. Thus to strike the number 25, two blows would be given, and then after a pause five more. Numbers containing ciphers and those made up of figures repeated, as 22, 33, etc., are not used for the signal-boxes.

SUMMARY.

The wire through which a current flows is magnetic.

This magnetism appears much stronger when the wire is bent into a coil, or helix.

When a piece of soft iron is placed inside a coil and a current sent through the wire, magnetism is developed in the iron. A magnet made in this way is called an *electro-magnet.*

In a left-handed helix, the end of the iron where the current enters the helix becomes a north pole, and the end where it leaves the helix a south pole. In a right-handed helix, it is just the reverse.

If the core of an electro-magnet is of soft iron, the magnetism can be destroyed by breaking the current, and the poles can be reversed by changing the direction of the current. If the core is of steel, it retains its magnetism permanently. Steel magnets are now usually made by means of the current.

Electro-magnets are usually of the horseshoe form. They are much stronger than other magnets. (236, 237.)

Page's Rotating Apparatus depends in its action on the fact that like poles of magnets repel and unlike poles attract each other; and that the poles of an electro-magnet can be reversed by changing the direction of the current. It illustrates one of the ways in which the electric force may be made to do mechanical work.

In Foucault's Rheotome the current of a battery is made to open and close its own circuit and also the circuit of a second battery. Every time the circuit is thus closed and opened, magnetism is developed and destroyed in an electro-magnet, and by this means the electric force is made to work a beam. This instrument, then, illustrates another mode of employing electricity as a motive power. (239.)

The electric force has also been used for regulating the motion of clocks. The pendulum of the regulating clock is made to play the part of a self-acting rheotrope. The pendulum ball of each copying clock is a helix with similar poles of permanent magnets facing it on each side. Thus the pendulum is made to vibrate every time the direction of the current is changed; that is, as often as the regulating pendulum vibrates. (240.)

Morse's Telegraph depends on the power of the current to develop magnetism, and hence is called an *electro-magnetic* telegraph. (241.)

The main battery is made to work a relay magnet, and a local battery to work the receiving instrument. By means of the relay magnet the operator can open and close the circuit of the local battery at a distance. (248.)

The Combination Printing Telegraph also depends on the power of the current to develop magnetism, and is, therefore, another form of electro-magnetic telegraph. The composing cylinder, type-wheel, and press are all driven by clock-work, the current being used only to remove the detent at the proper time to cause the press to print the required letter. (250.)

The Electric Fire-Alarm is another form of the electro-magnetic telegraph. The number of the signal-box near which the fire breaks out is telegraphed to the central station by turning a crank inside the box. The number of the box is then rung upon bells in various parts of the city by means of clock-work which is set in motion by electro-magnets. These magnets are included in a circuit which can be closed at the central station. (251.)

ELECTROLYSIS.

252. We have already seen that water and muriatic acid may be decomposed by the electric current (47, 60). The decomposition of a substance by electricity is called *electrolysis.* The literal meaning of the word is *loosening by electricity.* The substance decomposed by the electricity is called the *electrolyte.* The terminations of the metallic conductors through which the current passes into and out of the electrolyte are called *electrodes* (*roads of electricity*). That through which the electricity passes into the electrolyte is termed the *anode* (*road up*), and that through which the current passes out is termed the *cathode* (*road down*). The electrolyte is always decomposed into two parts, one of which appears at the anode and the other at the cathode. The former is called the *anion* (*going up*, or *to the anode*); the latter, the *cation* (*going down*, or *to the cathode*).

Thus in the experiment (47) of decomposing water in the U-tube, the water would be called the *electrolyte;* the decomposition of the water, *electrolysis;* the pieces of platinum connected with the battery, *electrodes;* the one connected with the positive pole, the *anode;* the one connected with the negative pole, the *cathode;* the oxygen which appears at the anode, the *anion;* and the hydrogen which appears at the cathode, the *cation.*

In the electrolysis of muriatic acid (60), chlorine is the *anion,* and hydrogen the *cation.*

Every chemical compound which is a conductor of electricity is an electrolyte when it is in the liquid state. Solid compounds are not decomposed by the current.

253. *The Electrolysis of Sulphate of Copper.*—If two electrodes of platinum are introduced into a solution of sulphate of copper (see Figure 86), bubbles of gas rise

from the anode. This gas may be collected by filling a test-tube with the solution of sulphate of copper, and inverting it over the anode. On testing the gas, it is found to be oxygen. On removing the cathode from the solution, it is found to be coated with copper. If one of the electrodes be of platinum and the other of copper, and the platinum be made the anode, the same results are obtained. If, however, the copper be made the anode, the cathode is still coated with copper, but no gas escapes from the anode.

Fig. 86.

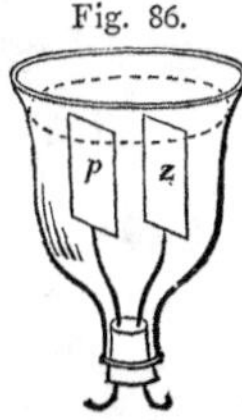

The most probable explanation of the above facts is as follows.

The electric current decomposes the sulphate of copper into copper and SO_4. This action may be expressed by an equation thus:—

$$CuSO_4 = Cu + SO_4.$$

Copper, appearing at the cathode, is the *cation;* and SO_4, appearing at the anode, is the *anion.*

When the anode is platinum, the anion acts upon the water of the solution, uniting with its hydrogen and setting its oxygen free.

$$H_2O + SO_4 = H_2SO_4 \text{ (hydrate of sulphuric acid)} + O.$$

So that the escape of the oxygen gas in this case is due to a secondary action, which is purely chemical.

When the anode is of copper, the anion, instead of acting upon the water, acts upon the anode itself.

$$Cu + SO_4 = CuSO_4.$$

Hence no oxygen escapes in this case, but sulphate of copper is formed as rapidly as it is decomposed; so that the solution always remains of the same strength. The anode is gradually eaten away and transferred to the cathode.

254. *Electrolysis of the Cyanides of Silver and of Gold.* — Analogous action takes place when the cyanide of silver or of gold (100, 115) is the electrolyte.

$$AgCy = Ag \text{ (cation)} + Cy \text{ (anion)}.$$

But when the anode is platinum,

$$H_2\Theta + 2\ Cy = 2\ HCy + \Theta;$$

and when the anode is silver,

$$Ag + Cy = AgCy.$$

In like manner,

$$AuCy_3 = Au \text{ (cation)} + Cy_3 \text{ (anion)}.$$

But when the anode is platinum,

$$3\ H_2\Theta + 6Cy = 6\ HCy + \Theta_3;$$

and when the anode is gold,

$$Au + Cy_3 = AuCy_3.$$

When any compound containing a metal is decomposed by electricity, the metal always appears at the cathode; and if the anode is of the same metal, the solution always remains of the same strength, while the anode is gradually transferred to the cathode.

255. *Electrotyping.* — When the solution of the sulphate of copper is decomposed slowly, the copper is deposited on the cathode in a coherent mass, which may be stripped off when it has become sufficiently thick. The sheet of copper stripped off is found to present a perfect reverse image of the face of the cathode, the faintest lines being copied with perfect distinctness. If this reverse image be now made the cathode, and another sheet of copper be deposited upon it, an exact copy of the original electrode is obtained. Any conducting substance, of whatever size and shape, may be made a cathode by simply connecting it with the negative pole of the battery. Hence coins,

medals, and engraved plates may be copied with perfect accuracy, and with but slight trouble and expense.

This process of copying by means of electricity is called *electrotyping*.

The face of a medal may be copied by making it the cathode and depositing a sheet of copper upon it, and then depositing another sheet of copper upon this sheet after it has been separated from the medal. In practice, however, a mould of the thing to be copied is first taken in some soft substance, such as plaster, gutta-percha, or wax, and this mould is made the cathode. If the mould is made of non-conducting material, as is usually the case, its surface must be covered with some conducting substance. The best material for this purpose is powdered graphite. The surface of the mould may be covered by means of a hair-brush with a film of graphite sufficient to make it a conductor, without obliterating the finest lines.

This copying by electricity is of the greatest importance in the arts. One of its chief uses is in copying printer's type after it has been set up, and in copying wood engravings. An impression is taken of the type or of the engraving in wax. This wax is then brushed over with powdered graphite, and made the cathode; the electrolyte is sulphate of copper, and the anode a piece of copper.

A large bath is used (see Figure 87), so that several

Fig. 87.

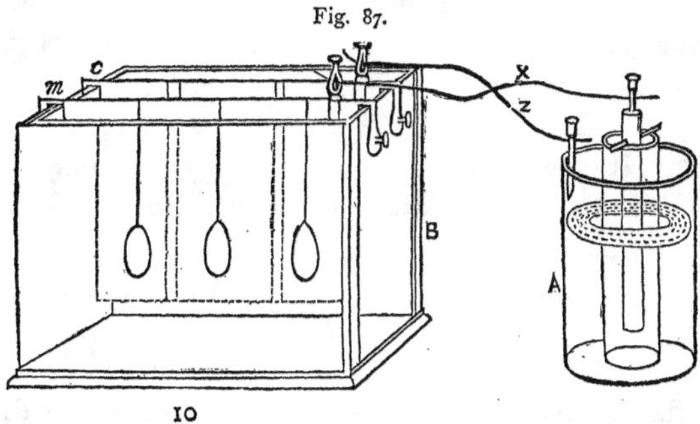

pieces may be electrotyped at the same time. These are all hung by wires to a metallic rod which is connected with the negative pole of the battery. Upon another metallic rod pieces of copper are hung opposite to the pieces to be copied.

A separate battery for generating the electricity is not absolutely necessary, though usually employed. The electric current may be generated in the bath itself. The object to be coated may be made to serve as one of the plates of the battery, and a piece of zinc as the other, the wire connecting the two being coated with insulating varnish.

256. *Electro-plating.*—This is the art of coating the baser metals with silver by the electric current. Articles to be electro-plated are generally made of brass, bronze, copper, or nickel silver, this last being the best material. Britannia metal, iron, zinc, and lead must be first electro-coppered, in order to be electro-plated, as silver does not adhere to the bare surfaces of these metals.

The articles to be plated are first boiled in caustic potash to remove any grease adhering to them; they are then immersed in dilute nitric acid to dissolve any rust or oxide that may have formed upon them; and lastly they are scoured with fine sand. Before being put into the silvering bath they are washed with nitrate of mercury, which leaves a thin film of mercury upon them, and this acts as a cement between the article and the silver. The bath is a large trough of earthenware or other non-conducting substance. It contains a weak solution of cyanide of silver in cyanide of potassium (water, 100 parts; cyanide of potassium, 10 parts; cyanide of silver, 1 part). A plate of silver forms the anode; and the articles to be plated, hung by wires to a metal rod lying across the trough, constitute the cathode. When the former is connected with the positive pole of a battery, and the latter with the negative pole, the silver of the cyanide begins to deposit itself on the suspended articles, and the cyanogen, set free at the plate, dissolves it, forming cyanide of silver. The reaction has already been explained (254). The thickness of the plating depends on the length of time the articles are immersed. The dull white appearance which they have when first taken from the bath is removed by polishing and burnishing.

257. *Electro-gilding.* — This process is essentially the same as electro-plating, except that the articles are coated with gold instead of silver. The electrolyte in this case is the cyanide or some other salt of gold, and the anode is a lump of gold. If it is not intended to gild the whole surface of the article, the parts not to be gilded must be coated with some non-conducting substance.

258. *Electro-metallurgy.* — Many other metals besides copper, silver, and gold may be deposited by electrolysis. The art of depositing, by electro-chemical action, a metal on any surface prepared to receive it, is called electro-metallurgy. All processes of the kind may be classified in two divisions, one of which is illustrated by electrotyping, and the other by electro-plating. The former includes all those cases in which the coating of metal has to be removed from the surface on which it is deposited; and the latter all cases where the coating remains permanently fixed. Gold, platinum, silver, copper, zinc, tin, lead, cobalt, and nickel can be deposited by electrolysis.

259. *Bain's Electro-chemical Telegraph.* — In 1846 Alexander Bain, of Edinburgh, constructed a telegraph which depends on the power of the electric current to decompose chemical compounds, and which is therefore called the *electro-chemical* telegraph.

When prussiate of potash is decomposed by the current, it is separated into prussic acid, which appears at the anode, and potash, which appears at the cathode. If the anode is of iron, the prussic acid acts upon it and forms the prussiate of iron, which is a blue salt.

If a sheet of common writing-paper be soaked in a solution consisting of six parts of prussiate of potash dissolved in water, two parts of nitric acid, and two of ammonia, the solution will scarcely color the paper. If now the paper, while still moist, be spread upon a metallic plate, which is connected with the negative pole of the battery, and an iron point connected with the positive pole be drawn over the paper, it will make a blue mark. If the point be disconnected with the battery, and then drawn over the paper, it will leave no mark. If a rheotome be introduced into the circuit between the iron point and the battery, and the paper be made to move uniformly under the iron point, a blue mark will be made on the paper while the current is flowing. If the current is on but an instant, a dot will be made; if for a longer time, a dash. Out of these dots and dashes an alphabet may be arranged, as we have already seen (242).

In the electro-chemical telegraph the sending instrument is a simple rheotome, like the one used with the Morse telegraph. (See Figure 79 and the accompanying description.)

The receiving instrument also is similar in its general construction to that of the Morse telegraph. The prepared paper is carried along by clock-work under the iron point, which marks it with dots and dashes, in the manner already described.

260. *The Voltameter.*—This instrument was invented by Faraday for testing the strength of a current. It is shown in Figure 88. Two platinum plates, each about half a square inch in size, are placed in a bottle containing water acidulated with sulphuric acid; the plates are soldered to wires which pass up through the cork of the bottle and terminate in binding screws; a glass tube fixed into the

cork serves to discharge the gas formed within. When the binding screws are connected with the poles of a battery, the water in the bottle begins to be decomposed, and hydrogen and oxygen are set free. If now the outer end of the discharging tube be placed in a trough of mercury, and a small graduated bell-glass, likewise filled with mercury, be placed over it, the combined gases rise into the bell-glass. *The quantity of gas given off in a given time measures the strength of the current.* The unit current may be taken as one which is capable of setting free one cubic centimetre of gas per minute.

Fig. 88.

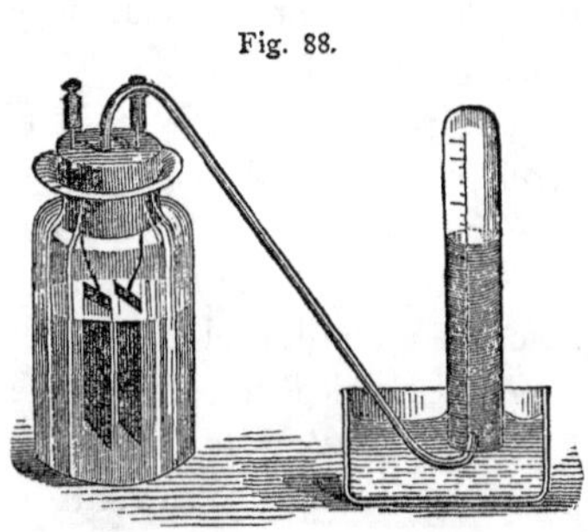

SUMMARY.

The electric current has power to decompose certain chemical compounds. These compounds must be in the liquid state, and must of course be conductors of electricity. Decomposition by means of the current is called *electrolysis;* the substance decomposed is the *electrolyte;* the conductors through which the current enters and leaves the liquid are the *electrodes,* the former being the *anode* and the latter the *cathode;* and the parts into which the electrolyte is separated, the *anion* and the *cation.* (252.)

When the electrolyte contains a metal this is always the cation, and is often deposited upon the cathode. Advantage is taken of this fact in electrotyping, electro-plating, and electro-gilding. (253 – 257.)

In electrotyping the metal deposited on the cathode is afterwards removed; in electro-plating and electro-gilding it is made to adhere permanently to the surface on which it is deposited. (258.)

Electrolysis is usually attended by a secondary action of a purely chemical nature. (253.)

The marks made on the prepared paper by Bain's telegraph are the results of electrolysis, or rather of the secondary action attending electrolysis. Hence this telegraph is an electro-chemical telegraph. (259.)

In the *voltameter*, electrolysis is applied to the measurement of the strength of currents. (260.)

THE POWER OF THE CURRENT TO DEVELOP HEAT AND LIGHT.

261. *Heat developed by the Current.* — When a current passes through fine wire, an intense heat is produced, sufficient in some cases to bring them to a white heat, and even to fuse platinum wire. Experiments upon the heating effects of the current may be made by the apparatus shown in Figure 89. The bottle is filled with alcohol, which is a non-conductor. The thick wires n and p are connected with a battery, and within the bottle they are joined with a fine spiral wire, surrounding the bulb of a delicate thermometer, t. When the circuit is closed, the heat developed is communicated to the alcohol, and thus to the thermometer. It is found that if the wire be kept the same, or of the same resistance, the heat is in proportion to the square of the strength of the current. Thus if a current of a certain strength raise

Fig. 89.

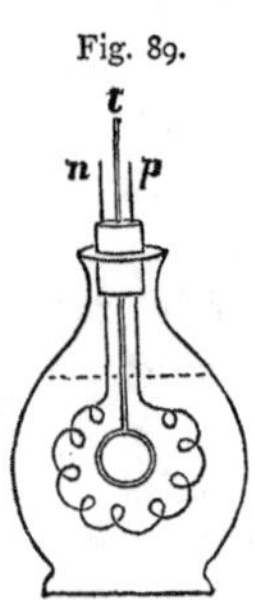

the thermometer 1° in a minute, a current of twice the strength will raise it 4° in a minute.

Again, if by means of a rheostat (235) the strength of the current be kept at the same point, and wires of different resistance be put into the bottle, the heat developed is in proportion to the resistance of the wire. Thus, if with a wire of a certain resistance the thermometer be raised 1° per minute, it will be raised 2° per minute with a wire of double the resistance.

Hence *the heat developed in a conducting wire by an electric current is proportional to the squares of the strengths of the current, and to the resistance offered by the wire.*

A very pretty illustration of the fact that the heat is proportional to the resistance is furnished by a chain, the alternate links of which are made of silver and platinum. When a current of sufficient strength is sent through the chain, the silver links remain black, while the platinum links become red-hot.

262. *The Current may be used to explode Gunpowder at a Distance.* — The power of the current to ignite fine wires of comparatively bad conducting metals, such as steel and platinum, is applied to the exploding of gunpowder at a distance. The current is transmitted to the point where the explosion is to take place by good conducting wires, and the fine wire is made to connect the ends of these wires in the gunpowder. A form of cartridge used for the purpose is shown in Figure 90. A tin tube is filled with gunpowder and stopped with a cork at each end. Two copper wires, insulated by being wound with woollen yarn, pass through one of the corks, and are connected within the tube by a fine steel wire soldered to each of them. When the current is sent through the wires, the explosion immediately follows. The fine steel wire is burnt away by the current,

Fig. 90.

but the copper wires are uninjured, and ready to be fitted up as before. Such cartridges are extensively used in blasting and mining. In long circuits the earth may serve as the return wire, as in the electric telegraph.

263. *The Electric Light.* — When the ends of two wires which form the poles of a powerful battery are made to touch, and then are separated for a short distance, the current does not cease with the separation, but forces its way through the intervening air, with an intense evolution of light and heat. The heat is sufficient to melt the most refractory metals, and therefore some substance rivalling the metals in conducting power, but much more infusible, must be found to act as the poles under such circumstances. The various forms of carbon are well suited to this purpose; but the best, both for conducting power and durability, is the coke formed in the retorts in the distillation of coal-gas. Figure 91 represents a simple arrangement

Fig. 91.

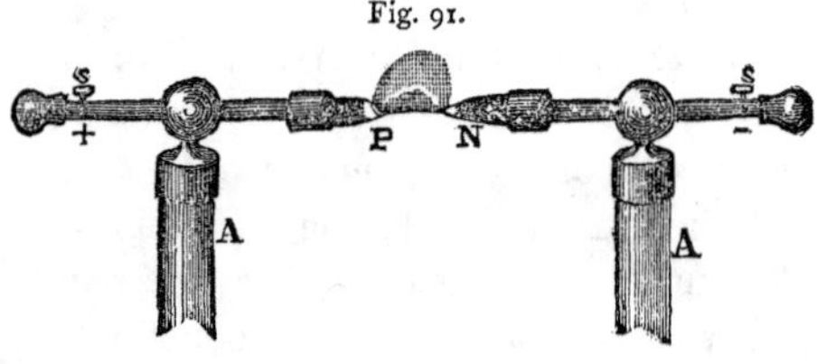

for producing the electric light. The carbon points, *P*, *N*, are fixed into hollow brass rods, which slide in the heads of the brass pillars, *A*, *A*, and are connected with the battery by binding screws, *s*, *s*. The points are made to touch, and the current is sent through the rods; the points are then separated a little, when a light appears between them rivalling that of the sun in purity and splendor. On examination this light is found to arise chiefly from the intense whiteness of the tips of the carbon points, and partially from an arch of flame extending from one to the other. The positive pole is the brightest and hottest, as is shown by the fact that, on intercepting the current, it continues to glow for some time after the negative pole has become dark.

While the light is kept up, a visible change takes place in the condition of the poles. The positive pole suffers a loss of matter; particles of carbon pass from it to the negative pole, which they partly reach, and partly are burned by the oxygen of the air on the way. The same takes place, though to a much less extent, with the negative pole. The positive pole becomes hollowed or blunted, and the negative remains pointed. This passage of matter between the poles accounts for the passing of the current through the air, as thereby a conducting medium extends between the poles.

The heat of this arch of flame, or *voltaic arc*, as it is called, is the most intense that can be produced. Platinum melts in it like wax in the flame of a candle. Quartz, the sapphire, magnesia, lime, and other substances equally refractory, are readily fused by it. The diamond becomes white hot, swells up, fuses, and is reduced to a black mass resembling coke.

The electric light is caused, not by the combustion of the carbon, but by its incandescence. The light can consequently be produced in a vacuum, and below the surface of water, oils, and other non-conducting liquids. It is thus quite independent of the action of the air.

With a battery of some fifty Bunsen's elements, a light is produced of very great brilliancy; but when very great power is to be obtained, twice or thrice that number must be employed.

264. *Electric Lamps.* — Various arrangements have been invented for giving steadiness to the electric light by keeping the carbon points within such a distance of each other that the current can pass between them. Foucault, aided

by Duboscq, was the first (1849) to construct an electric lamp of this description. In it, by aid partly of an electro-magnet, and partly of clock-work, the two points are made to travel towards each at rates corresponding to those of their combustion, the positive pole moving faster than the negative. A detent is fixed to the keeper of the electro-magnet which stops the clock-work when the keeper is drawn to the magnet, and starts it when the keeper is released. A spring carries the keeper away from the magnet when the current does not circulate or becomes very weak. Thus, when the carbon-points waste away and separate from each other, the current is weakened, and when it gets so weak as to impair the splendor of the light, the spring carries away the keeper and thereby starts the clock-work. The points now approach each other until the current becomes strong enough to cause the magnet to draw the keeper to itself and stop the clock-work. The points are thus kept at just the distance for producing the most brilliant light.

The electric light has not yet been used successfully for lighting streets and public places. One difficulty is found in the uncertainty of the light and the care attending its use. By means of the electric lamp just described the light may be continued for hours, but even then it is not perfectly steady, and the apparatus cannot be safely left without an attendant. Another difficulty arises from the striking and unpleasant contrast of light and shadow accompanying it, rendering, as it were, the surrounding gloom as manifest as the brightness of the light itself. It has, however, been used with excellent effect where a limited space had to be lit up for a few nights, as in the construction of bridges and the like. It has also been applied with success to light-house illumination. The light-house at Dungeness, on the coast of Kent, England, has been lit up with it since 1862, and that at La Heve, near Havre, since 1863. It has been found from these that the power of the electric light to penetrate fogs is immensely superior to that of the usual oil light.

265. *The Electro-thermal Telegraph.*—If a fine platinum wire introduced into the circuit be bent, and a piece of paper be drawn over the point while the current is passing, a line will be burned in the paper. If the paper is drawn along at a uniform rate by means of clock-work, and the point rests upon the paper, a dot will be burned into it when the current passes but an instant; and a dash when it passes for a longer time. Out of these dots and dashes an alphabet may be made similar to that devised by Morse (242). Hence we see that a telegraph can be constructed depending on the power of the current to develop heat. Such a telegraph has been invented by Horne, and is called the *electro-thermal* (that is *electric-heat*) *telegraph.* It cannot, however, be worked without a local battery and a relay magnet (248). These are specified in Morse's patent as his invention; and the instrument cannot therefore be worked without the permission of the owners of that patent. It has no point of superiority over the Morse telegraph, and, as it cannot be used as a substitute, it has never been put into practical operation.

SUMMARY.

When the current passes through a conductor, heat is developed. When the conductor is a fine wire, the heat often becomes very intense. The heat is found to be proportional to the square of the strength of the current, and to the resistance offered by the conductor. (261.)

By introducing a poor conductor into any part of the circuit, heat may be developed at that point. Advantage is taken of this fact in blasting under water and at a distance. The circuit is completed through the gunpowder by means of a very fine wire, which becomes intensely heated when the current passes, and thus ignites the powder. (262.)

Advantage is taken of the same fact in the development of the electric light. Carbon points are introduced into the circuit, and after the current begins to pass they are separated a little way, and the space between them becomes filled with carbon particles. These constitute an imperfect conductor, and the resistance encountered by the current in passing through it develops intense heat and light. (263, 264.)

The power of the current to develop heat has been applied to telegraphy in the invention of the *electro-thermal telegraph*. (265.)

VOLTAIC ELECTRICITY.

266. We have found that an electric current can be obtained by connecting the two poles of a battery by means of a copper wire or other conductor. (222.)

What now is the origin of this electric force?

In seeking an answer to this question, it is desirable to have a simpler battery than the one already described. If a strip of amalgamated zinc* (zinc which has been immersed in mercury) and another of copper are placed in a cup of dilute sulphuric acid, and the two strips are joined by means of a copper wire (see Figure 92), a current is obtained. We have here a battery of very simple construction, consisting of only one liquid and two metals. So long as the plates are unconnected by the wire, they

* See Appendix, 32.

are unaffected. The moment they are connected, bubbles of hydrogen gas appear in abundance at the copper plate. If the connection continues for some time, and the plates and the liquid are then examined, the copper plate is found to weigh exactly the same as at first; while the zinc plate weighs less, and the liquid is found to contain sulphate of zinc in solution. If dilute sulphuric acid were put into a glass vessel, and a current sent through it, the acid would be decomposed into hydrogen and $S\Theta_4$; and if the anode were of zinc, the $S\Theta_4$ would unite with it, forming sulphate of zinc. We thus see that the action inside the cell is a case of electrolysis attended by the usual secondary action (253).

Fig. 92.

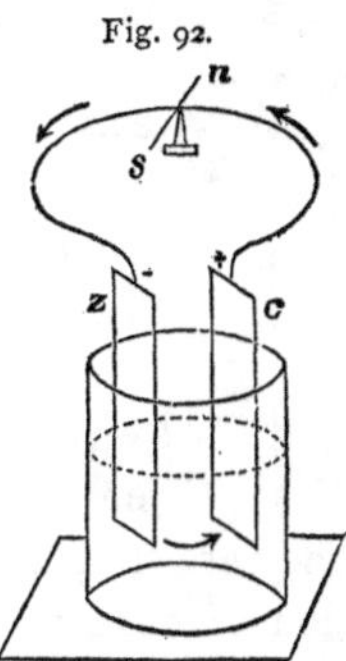

If the plates are so arranged as to be in the plane of the meridian, and a small magnetic needle is held near the surface of the liquid when the plates are connected, the needle will be deflected in such a manner as to show that a current is passing through the liquid from the zinc to the copper. If the same needle is held over the wire which connects the plates, it is deflected in the opposite direction, showing that the current is passing through the wire from the copper to the zinc.

Two plates connected as above described constitute a *galvanic* or *voltaic pair.*

267. *The Contact Theory and the Chemical Theory.* — Since the current appears only when the wires from the two plates are brought in contact, many scientific men hold that the electric current has its origin in the contact of dissimilar metals, and that the chemical action is the result of the current thus originated. This theory is known as the *contact theory.*

But an electric current can be developed in a voltaic

pair without contact of dissimilar metals. This may be proved by the simple experiment shown in Figure 93. *z* is a plate of zinc bent at right angles; *p*, a platinum plate, to which a platinum wire is attached. At *a* a small piece of blotting-paper, moistened with a solution of starch and iodide of potassium, is placed between the plate of zinc which supports it and the platinum wire which rests upon it. No change occurs in the solution of the iodide until the plates are immersed in dilute nitric acid; but very soon after they are thus immersed, evidence of a current in the direction of the arrow is afforded by the appearance of a blue spot where the wire touches the paper. The iodide of potassium is decomposed by the voltaic action, and the iodine unites with the starch to form the blue compound.

Fig. 93.

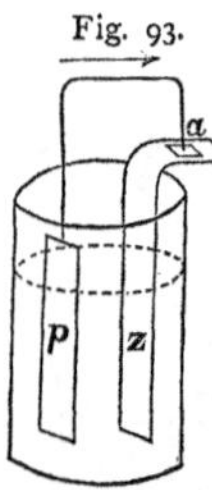

On the other hand, without chemical action there is no current. Let an iron wire be connected with one extremity of a galvanometer of moderate sensibility, and a platinum wire with the other extremity. Immerse the ends of the wires in concentrated nitric acid, without allowing them to touch each other in the liquid; and there will be no chemical action upon the iron, and no movement of the magnetic needle. But the addition of a little water will cause a rapid solution of the iron in the acid, and the needle, the moment the chemical action begins, will be strongly deflected.

Again, let two glass vessels be filled with sulphide of potassium. Put two platinum plates into one, and a platinum and an iron plate into the other. Connect one of the platinum plates in each vessel with a galvanometer. To the other platinum plate attach a platinum wire, and to the iron plate an iron wire. The sulphide of potassium is a conductor of electricity, but is chemically inactive when associated with iron and platinum in a circuit. When the platinum and iron wires are joined, if the electric force were developed by their contact, a current would be generated, since we have all the conditions necessary for a circuit. But the galvanometer gives not the slightest evidence of a current. If zinc be interposed between the ends of the platinum and iron wire, the result is the same; but if a piece of paper moistened with sulphuric acid be placed between the ends of the wires, the needle is deflected at once. We have here conclusive evidence that the simple contact of the iron and the platinum does not develop electric force, and that this is developed only by the chemical action of the acid upon the iron.

Into a glass vessel containing sulphide of potassium put two plates, one of copper and the other of silver, and connect them with the galvanometer. The needle at first turns in a direction which shows that the copper is the positive pole; it then gradually returns to its first position, and is again deflected in the opposite direction, showing that the silver is now the positive pole. After some time it returns and is deflected in the original direction, and goes on thus changing. If the plates be examined during these changes, it is observed that sulphide of copper is formed when the copper is the positive pole, and sulphide of silver when the silver is the positive pole. The alternate action is owing to the relative condition of the two plates when coated with their sulphides. The electric current of a voltaic pair is thus shown to be not invariable in direction, as it would be according to the contact theory, but to change its direction with the seat of chemical action. It always starts from the plate acted upon, passes through the liquid to the passive plate, then through the wire back again to the active plate.

Therefore it is now generally held that the current originates in the chemical action, though it cannot appear until a road for it has been formed by connecting the plates by a conductor outside the liquid.

When the amalgamated zinc is put into the cell (265), the $S\Theta_4$ of the acid acts upon it, not strongly enough to effect combination, but sufficiently to develop a current, which, in passing through the cell, decomposes the acid into hydrogen and $S\Theta_4$, leaving the latter free to combine with the zinc. Its combination with the zinc develops more electricity, and the process thus goes on. This is known as the *chemical theory*.

It is evident from what has just been stated, that a voltaic pair cannot be constructed by using one liquid and two plates of the same metal. If both plates were equally acted upon, currents of equal strength would start from each in opposite directions, and would exactly neutralize each other. On the other hand, if one plate is acted upon very strongly, and the other not at all, there will be no current from the latter to interfere with that from the former. Hence platinum and zinc constitute a better voltaic pair than copper and zinc, when the acid used is sulphuric acid. In the former case, the acid does not act upon the platinum at all, and no counter-current is started; but in the latter it tends to act upon the copper to some extent, as well as upon the zinc, and a counter-current is established which weakens the original current.

Fig. 94.

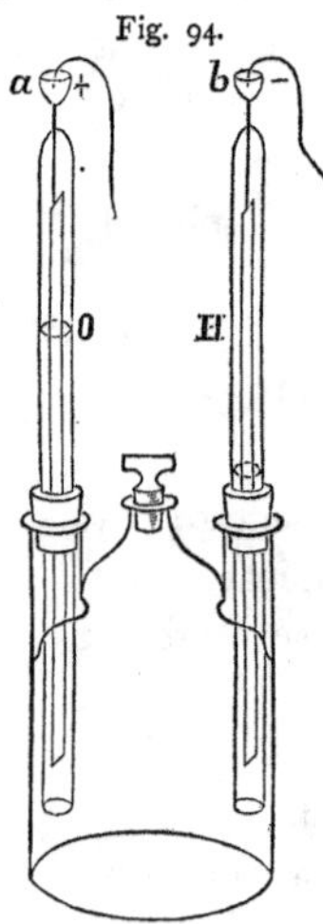

268. *Grove's Gas Battery.*—Into the two outer necks of a three-necked bottle (see Figure 94) two glass tubes, *O*, *H*, are fitted by means of corks. Each of these tubes is open below, and contains a long strip of platinum, from which a platinum wire passes through the glass at the top. The wires terminate in little cups, *A*, *B*, containing mercury. The whole

apparatus is filled with acidulated water, and the poles of a galvanic battery are placed in the little cups. Water is thereby decomposed; oxygen collects in one tube, and hydrogen in the other. When the battery wires are removed, no change takes place till the cups are connected by a wire. The oxygen and hydrogen then gradually disappear, and an electric current passes from the oxygen to the hydrogen.

The operation of this cell shows that, when the ions are allowed to collect about the plates, they tend to send a current in the opposite direction to that passing in the cell.

269. *Grove's Battery.* — But we have already seen that the chemical action within the cell of an ordinary battery is a case of electrolysis. The acid commonly used to act upon one of the plates is sulphuric acid, and when the cell is in action bubbles of hydrogen are seen to collect upon the passive plate so as nearly to cover it. This interferes with the action of the cells in two ways: (1.) it prevents contact between the plate and the liquid; (2.) the hydrogen tends to send a counter-current through the liquid, as is shown in the case of Grove's gas battery, and this counter-current of course partially neutralizes the original current.

The ion does not tend to collect about the active plate, because it immediately combines with it, forming a soluble compound which passes into the solution.

It therefore becomes desirable to prevent the collection of the hydrogen about the passive plate. This is done in Bunsen's battery by surrounding the passive plate with strong nitric acid, which takes up the hydrogen. The principle of this battery is due to Grove, who constructed a battery of which Bunsen's is merely a modification. In Grove's battery (shown in Figure 95) a strip

Fig. 95.

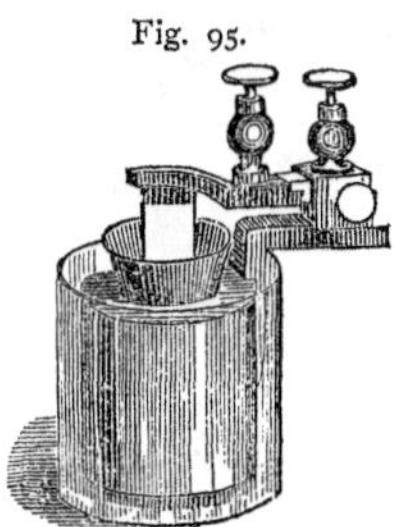

of platinum is used for the passive plate, instead of carbon. In other respects it is essentially the same as Bunsen's.

270. *Daniell's Battery.* — A cell of this battery is shown in Figure 96. The outer vessel is of copper, and serves as the passive plate. Inside this is a vessel of porous earthen-ware, containing a rod of zinc. The space between the copper and the porous cup is filled with a solution of sulphate of copper, which is kept saturated by crystals of the salt lying on a perforated shelf. The porous cup is filled with dilute sulphuric acid. The porous partition keeps the fluids from mingling, but does not hinder the passage of the current. The sulphate of copper which is in contact with the passive plate serves to take up the hydrogen. There are two other reasons for putting the sulphuric acid within the porous cup: (1.) if the sulphuric acid came in contact with the copper, it would tend to act upon it as well as upon the zinc, and thus to send a counter-current through the cell; (2.) the sulphate of zinc formed is thus kept from coming in contact with the copper. If it were allowed to come in contact with the copper, it would be decomposed by the current passing through the cell, and zinc would be deposited on the copper; and both plates would soon be virtually of the same metal.

Fig. 96.

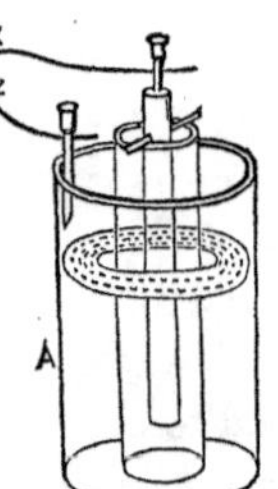

SUMMARY.

A voltaic pair consists of two dissimilar conducting plates, usually of metal, immersed in a liquid which will act chemically upon one of them. The plate acted upon is called the *active* plate; the other, the *passive* plate. The liquid used must always be an electrolyte; and the chemical action going on in the cell is electrolysis of the liquid, attended by the usual secondary action. (266.)

According to the *contact theory*, the current originates in the contact of dissimilar metals. But no current can be developed by such contact without chemical action, even though the liquid used is a good electrolyte; while a current can be developed by chemical action without the contact of dissimilar metals, and even by using only one metal. Moreover, the strength of the current depends upon the intensity of the chemical action, and its direction upon the seat of this action; the current always starting from the active plate and passing through the liquid to the passive plate, and then through the wire to the active plate again. If by any means the seat of chemical action changes, the direction of the current also changes. These facts show that the current must originate in the chemical action between the liquid and the active plate. This chemical action may not be strong enough to effect combination until a current has been started, which helps to decompose the liquid, and thus aids the chemical action at the active plate. (267.)

When the ions are allowed to collect about the plates, they tend to send a current in the opposite direction to the one passing through the liquid.

In all ordinary batteries, the active plate is zinc, and the liquid acting upon it is dilute sulphuric acid. No ion tends to collect on the active plate, since the ion which appears there enters at once into combination with the plate, giving rise to a soluble compound which passes into solution. The ion hydrogen which appears at the passive plate collects about it.

This ion interferes with the action of the cell by preventing the plate from coming thoroughly in contact with the liquid, and by tending to send a current in the opposite direction to the one originated by the active plate.

In Grove's and Bunsen's batteries, the collection of this ion about the passive plate is prevented by immersing the plate in strong nitric acid contained in a porous cup. The nitric acid takes up the hydrogen. (268, 269.)

In Daniell's battery, the plates are zinc and copper; the former immersed in dilute sulphuric acid, the latter in a solution of sulphate of copper. The solution of sulphate of copper takes up the hydrogen, and thus prevents it from collecting on the copper plate. The sulphuric acid is contained in a porous cup, which prevents it from coming in contact with the copper, and thus generating a counter-current; and the sulphate of zinc formed is also kept in the porous cup, so that no zinc can be deposited on the passive plate. (270.)

Platinum and zinc constitute a more effective voltaic pair with dilute sulphuric acid than copper and zinc; since the acid has no tendency to act upon the platinum so as to start a counter-current, while it has a tendency thus to act upon the copper. A voltaic pair cannot be constructed by using two plates of the same metal and one liquid, since the counter-current would then just equal the direct current, and thus neutralize it. (267.)

MAGNETO-ELECTRICITY.

271. *Electricity developed by a Magnet.* — We have now seen that the electric current has power to decompose chemical compounds under certain conditions; and also that chemical action under certain conditions has power to originate the electric current.

We have also seen that the electric current has power to move a bar magnet.

If the two ends of a copper wire are connected with a delicate galvanometer, and then the wire is carried quickly over the pole of a powerful magnet, or, what amounts to the same thing, the pole of the magnet is moved quickly by the wire, the needle of the galvanometer will be deflected; showing that a magnet moving about a wire, or a wire about a magnet, has power to originate an electric current.

A current is generated in a continuous conductor, whenever it is made to cut the lines of magnetic force (214), either by moving it about a magnet, or a magnet about it. The direction of the current depends upon the direction in which the wire is moved across the curves. When the wire is made to cut the curves in opposite directions, the current flows in opposite directions. A current thus originated by a magnet is said to be *induced* by it, and the electric force thus induced is called *magneto-electricity.*

It has been shown (236) that the electric current, when it passes through a helix, renders a piece of soft iron placed inside the helix temporarily magnetic.

If the two ends of a helix are connected with a delicate galvanometer, and a piece of soft iron is placed inside the helix, and the north pole of a magnet is brought against one end of the iron, and the south pole of another magnet against the other end, the soft iron becomes magnetic and the galvanometer needle is deflected; showing that the soft iron in becoming magnetic generates a current in the wire of the helix. The needle will soon return to its original position. If now the magnets are removed from the soft iron, the needle is deflected in the opposite direction; showing that the soft iron in losing its magnetism also originates an electric current in the wire of the helix, and that the direction of the current depends upon whether the wire is gaining or losing its magnetism.

It appears, then, that a current may be developed in a conductor by using either a constant or a variable magnet. When a constant magnet is used, the current is developed by changing the relative positions of the magnet and the conductor; when a variable magnet is used, by changing the strength of its magnetism.

272. *Magneto-Electric Machines.*—An instrument for developing electricity by means of magnetism is called a *magneto-electric machine.* In ordinary machines of this kind the electricity is induced by means of a variable magnet (271): there must, therefore, be some means of developing and destroying magnetism in a piece of soft iron. The iron is placed inside a helix, which serves as a conductor for the current induced. The magnetism may be developed and destroyed by means either of a permanent magnet or of an electric current.

The former method is illustrated by Figure 97. *NS* is a permanent horseshoe magnet. *CD* is a bar of soft iron with coils, *A* and *B*, wound round its ends, and may be viewed as the armature of the magnet. *C D* is capable of rotation round the axis *E F.* So long as *CD* remains at rest, no currents are induced in the coils, for no change takes place in the magnetism induced in it by the action of *NS.* But if the poles of *CD* leave *NS*, the magnetism of the soft iron diminishes as its distance from *N S* increases, and when it stands at right angles to its former position, the magnetism has disappeared. During the first quarter-revolution, therefore, the magnetism of the soft iron diminishes, and an

Fig. 97.

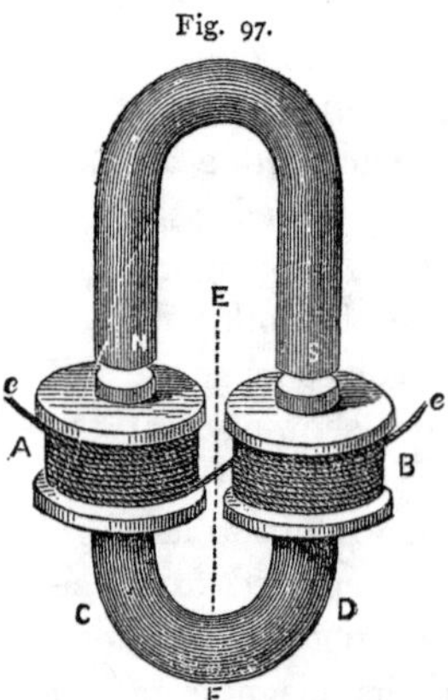

electric current is induced in the coils. During the second quarter-revolution the magnetism of the armature increases till it reaches a maximum when its poles are in a line with those of *N S.* A current also marks this increase, and moves in the same direction as before; for though the magnetism increases instead of diminishing, which of itself would reverse the induced current, the poles of the armature, having changed their position with relation to those of the permanent magnet, have also been reversed, and this double reversal leaves the current to move as before. For the second half-revolution the current also moves in one direction, but opposite to that of the first half-revolution, since the position of the armature is reversed. Thus *in one revolution of a soft iron armature in front of the poles of a permanent magnet, two currents are induced in the coils encircling it, each lasting half a revolution, starting from the line joining the poles.*

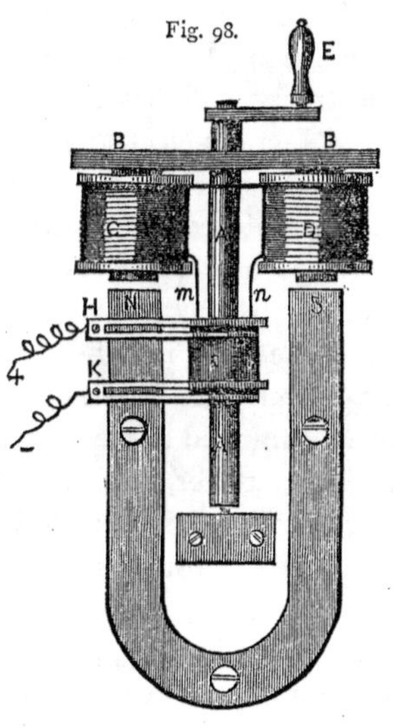

Fig. 98.

The manner in which the armature may be made to rotate, and the current to flow constantly in one direction, is shown in Figure 98, which represents a common form of magneto-electric machine. *N S* is a fixed permanent magnet. *B B* is a soft iron plate, to which are attached two cylinders of soft iron, round which the coils *C* and *D* are wound. *C B B D* is thus the revolving armature, corresponding to *C D* in Figure 97. *A A* is a brass rod attached to the armature, and serving as its axle. *F* is a cylinder fastened to *A*, and is pressed upon by two fork-like springs, *H* and

K, which are also the poles of the machine. The ends *m* and *n* of the coil are soldered to two metal rings on *F*, insulated from each other. When the armature revolves, *A A* and *F* move with it. *F*, *H*, and *K* are so constructed as to act as a rheotrope, reversing the current at each half-revolution. By this arrangement, the opposite currents proceeding from the coil at each half-revolution are so transmitted to *H* and *K*, that these retain their polarity unchanged. When the armature is made to revolve with sufficient rapidity, a very energetic and steady current is generated, which has all the properties of the galvanic current. Compared with the galvanic battery, the magneto-electric machine is a readier, steadier, and cleanlier source of electricity, and has come to be extensively used instead of it. Magneto-electric machines may be made of any strength by increasing the number of magnets and the mechanical force employed.

In large machines, several magnetic batteries are employed, with a corresponding number of armatures and coils. The coils may be arranged like the cells of a galvanic battery, for tension or for quantity. For tension, they are arranged successively, so that they form one compound circuit; for quantity, each coil or set of coils contributes to the common current. The thickness of wire is selected according to the object of the machine. For giving shocks, or for electrolysis, the wire must be long and fine; for heating platinum wire, thicker and shorter. The electric force increases with the rapidity of rotation.

273. *Wilde's Magneto-Electric Machine.* — A magneto-electric machine of great power has been recently invented by Mr. Wilde, of Manchester, England. A front view of the machine is shown in Figure 99. *M* is the foremost of a series of sixteen powerful steel magnets of horseshoe form placed one behind another in a horizontal row. These magnets are fixed below to the *magnet cylinder*,

Fig. 99.

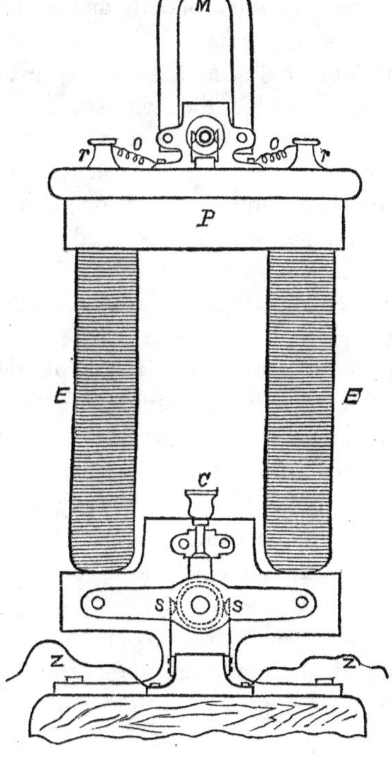

shown on a larger scale in Figure 100. This is made partly of iron, partly of brass. The sides *i i* are of iron, and the brass bars *b b* lie between them. In the centre is a circular hole extending the whole way through. The magnets are firmly fastened to the iron sides *i i*, so that the latter form the poles of the magnetic battery, the brass bars between them insulating them from each other.

A cylindrical armature *a a* of cast-iron is made to revolve within the magnet cylinder. Its diameter is a little less than that of the cylindrical hole, so that it can revolve without friction very close to the polar surfaces. It is shown in section in Figure 100. Two rectangular grooves are cut in it, as there represented, and in these about fifty feet of insulated copper wire is wound lengthwise in three coils. The coil thus formed is shut

Fig. 100.

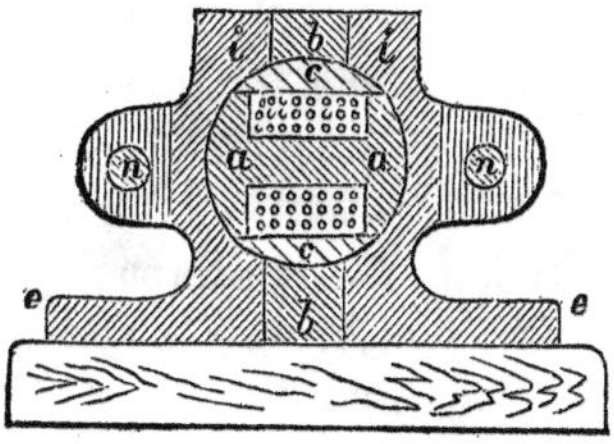

in by wooden packing, *c c*. Two caps of brass are fitted to the ends of the armature, and to these are attached the steel axes of rotation. The rear axis is connected by means of a pulley and belt with the engine which rotates the armature. On the front axis are two metallic pieces, one connected with the armature, and the other insulated from it. One end of the armature coil is connected with the armature, and thus with one of these metallic pieces, and the other end is insulated from the armature and connected with the other piece; so that these metallic pieces are the terminals of the coil. Two steel springs press against these pieces, each spring against one piece during half a rotation. In the position shown in Figure 100, the armature is magnetized, since the parts *a a* are facing the poles of the permanent magnets. On performing a quarter-revolution, the armature loses its magnetism, since its poles are carried away from the poles of the magnets. On accomplishing another quarter-revolution, it again becomes magnetic, and so on; so that in one revolution the armature induces two opposite currents in the coil, one in each half-revolution. The springs act as a rheotrope, and thus cause the current to pass through them always in the same direction. The armature is made to revolve some 2,500 times per minute, sending 5,000 waves or currents of electricity to the wires *o o*.

One advantage of the position of the armature in this machine is that its motion is not resisted by the air. In the ordinary magneto-electric machines (see Figure 98) much of the mechanical force applied to the rotation is wasted in beating the air.

Another advantage is that the inductive action of the magnet is exerted directly on the coil, as well as through the intervention of the armature. If the coil were made to rotate without the armature, currents would be induced in it of the same kind as that induced by the armature, though of feebler intensity; and these currents would be strongest when the coil was moving through the line joining the poles, and weakest when it was at right angles to that position. The currents induced by the armature are strongest when those just mentioned are weakest, and weakest when those are strongest; so that armature and coil combine to make the current uniform.

But the chief peculiarity and merit of Wilde's machine is, that the current got from the magneto-electric apparatus is not directly made use of, but is employed to magnetize an electro-magnet some hundreds of times more powerful than the magnetic battery originally employed, and this electro-magnet is made to induce another and proportionally more powerful current by means of a second rotating armature.

This electro-magnet, *E E* (Figure 99), forms the lower part of the machine, and by far the most bulky portion of it. The upper and lower machine are in action precisely alike; only the upper magnet is a permanent magnet, and the lower one an electro-magnet. We have the same magnet-cylinder, the same armature, springs, and poles. This armature is made to rotate some 1800 times per minute.

A machine intended for a three-horse power steam-engine, and worked with that power, will consume carbon sticks three eighths of an inch square, and evolve a light of surpassing brilliancy. With a machine consuming carbons half an inch square, the light is of sufficient intensity to cast shadows from the flames of street-lamps a quarter of a mile off. The same light, at two feet from the reflector, darkened photographic paper as much in twenty seconds as the direct rays of the sun at noon in one minute.

Wilde's machine enables us to convert any amount of mechanical force into electricity by increasing the size of the electro-magnet, or by using a second electro-magnet induced by the first; so that *a magnet indefinitely weak can be made to induce a current or a magnet of indefinite strength.* The size and weight of the apparatus are also small.

274. *Induction Coils.* — When the magnetism is developed and destroyed by means of a current (272), the soft iron must be placed inside a coil through which the current is sent. This is called the *primary coil,* and must be placed inside another coil, called the *secondary coil,* which serves as a conductor of the electricity developed by induction. A magneto-electric machine made in this way is commonly called an *induction coil.*

Fig. 101.

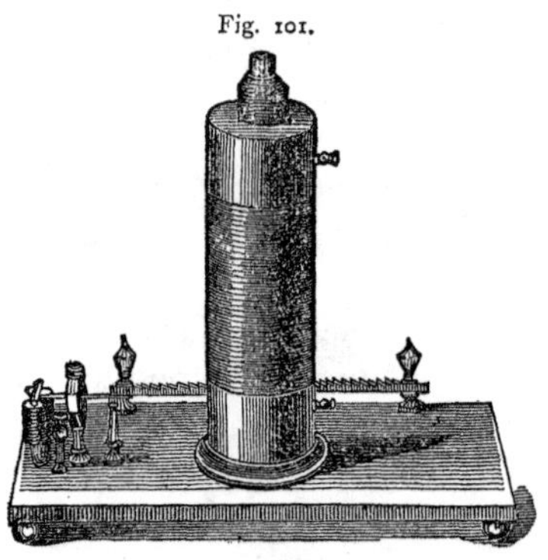

One of simple construction is shown in Figure 101. The primary coil is of coarse wire wound with wool, and is attached to the wooden base of the instrument. The secondary coil is of finer silk-wound wire, much longer than the primary wire. Within the primary coil is a bundle of iron wires, which are sufficiently insulated by the rust that gathers on them. The developing of magnetism in these wires is the chief aim of the primary coil, and, as a strong current is necessary for that purpose, coarse wire is used in that coil. In the secondary coil, the tension of the induced current alone is aimed at, and fine wire is used, so that as many turns as possible may be brought within the influence of the primary coil and its core; for it is found that *the tension of the induced current is proportional to the strength of the primary current and to the square of the resistance in the secondary coil.*

In order, however, to obtain the greatest effect from the

secondary coil, it is necessary to have some means of rapidly completing and breaking the primary current. This is effected in the instrument under consideration, either by means of the rasp seen behind the coils, or by the self-acting rheotome at the left hand. When the former is used, one of the battery-wires is attached to one of the binding-screws, and thereby to one end of the primary coil; and the other battery wire is drawn along the teeth of the file, which is connected with the other end of the coil. The current is stopped and started again every time the wire passes from one tooth to another; and every time it is stopped or started, an inverse or a direct current is excited in the secondary wire. The rheotome breaks the current in the same way, but more regularly and rapidly. When it is used, both battery wires are attached to the binding-screws, bringing the rheotome and the primary coil into the circuit.

275. *The Inductorium, or Ruhmkorff's Induction Coil.* — The essential parts of this apparatus are the same as in that shown in Figure 101, and described in the last section. A primary coil, with its core of iron wire, and a secondary coil outside the primary and insulated from it, form the main portion of the instrument. The primary coil is connected with the poles of a galvanic battery, and a rheotome is introduced into the circuit to effect the interruptions of the current essential to its inductive action. A rheotrope and a *condenser* are also connected with the primary circuit.

The best rheotome is Foucault's (239); the best rheotrope, the one described in section 227. These two are usually combined in one instrument.

The *condenser* consists of several sheets of tinfoil and silk, laid alternately upon one another. The first, third, fifth, etc. sheets of tinfoil are connected by strips of the same material; so are the second, fourth, sixth, etc. Each set of sheets is connected with one of the wires of the primary coil. The condenser is generally placed within the wooden base of the instrument, and does not meet the eye. The object of the condenser is to absorb the *extra current*, to be described hereafter. (279.)

The excellence of the instrument depends on the proper insulation of the secondary coil. The bobbin must be made of glass, gutta-percha, or vulcanite, which is best of all, so as to prevent the induced electricity from reaching the ground by the primary coil. Care must also be taken to insulate the different parts of the secondary coil from each other, so that the induced current may not shorten its course by leaping over one or more layers of the coil.

A Ruhmkorff's coil of moderate size readily yields sparks of from four to five inches, with a battery of six Bunsen's cells. The power of the induced current to deflect the needle of the galvanometer, and to effect electrolysis, is very insignificant. This shows that it is very much inferior to the inducing current in quantity, however much it may be superior in tension. The physiological effect, however, is tremendous, and the experimenter must take care not to allow any part of his body to form the medium of communication between the poles, as the shock might be dangerous, if not fatal.

276. *The Electric Egg.* — When the induced current is made to pass through highly rarefied air, a very beautiful effect is produced. This may be shown by the *electric egg*. It consists (see Figure 102) of a glass vessel with an open neck above and another below. The lower opening is fitted with a brass stopcock, and can be screwed to the plate of an air-pump. A brass rod and ball rise a little way into the egg. Another brass rod terminating in a ball slides air-tight through a cap covering the upper opening, so that the two balls can be set at different distances from

each other. When the air is exhausted from the egg, and the wires from the coil are attached above and below, a luminous glow extends between the balls. When the exhaustion is more complete, black bands are seen to cross the light horizontally, as shown in the figure. The cause of this appearance is not fully understood. The ball which forms the negative pole is enveloped with a blue light. The glow, which is of a beautiful mauve tint, appears to proceed from the positive ball, and reaches nearly to the negative ball, from which it is separated by a well-marked non-luminous space. If the direction of the primary current is changed, these appearances are instantly transposed.

Fig. 102.

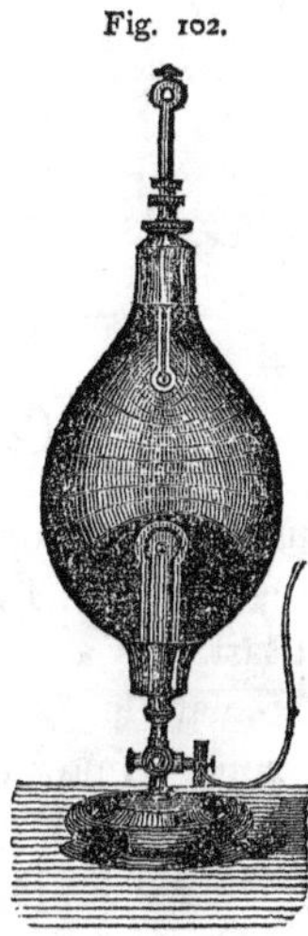

277. *Geissler's Tubes.* — A variety of forms of apparatus are used for showing the electric light in rarefied air and in other gases.* *Geissler's tubes*, so called from the inventor (a German, who alone knows the secret of their manufacture), are combinations of bulbs and tubes, filled with rarefied gases and liquids, and then sealed air-tight, so as to be ready for use at any time. One of them is shown in

Fig. 103.

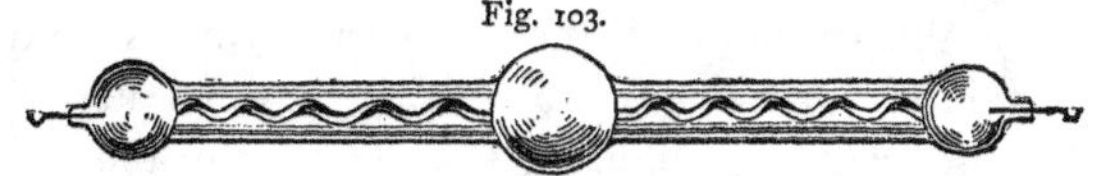

Figure 103. When the current is sent through these tubes, they exhibit lights of various tints according to the gases contained in them.

278. *Current Induction.* — Two long copper wires (see

* All the experiments with the electric light usually performed by means of Frictional Electricity can be better performed with the *Inductorium*. See Appendix, 31.

Figure 104) are fixed so as to be parallel and close to each other. The ends of the one, *p p*, are connected with the poles of a battery *E*, and those of the other, *s s*, with a galvanometer *G*. The instant the circuit of the battery is completed, and the current sent through *p p*,

Fig. 104.

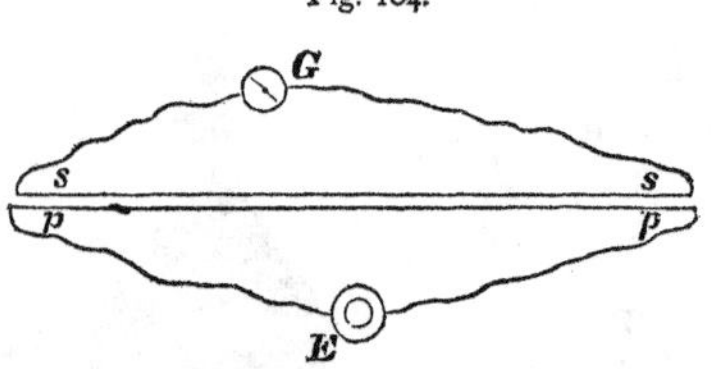

a current in the opposite direction is induced in the wire *s s*, as is shown by the galvanometer. This induced current lasts but a moment, for though the current continues to circulate in *p p*, the needle soon falls back to its original position. If now the battery current in *p p* be interrupted, another momentary current in *s s* is shown by the galvanometer, but in this case in the same direction as the inducing current. The inducing wire and current are called *primary;* the other wire and current, *secondary.* If the primary wire be movable, so that it can suddenly be brought near to and withdrawn from the secondary, while the battery current flows steadily, currents are induced as before, the approach of the wire being marked by an inverse current, and its withdrawal by a direct one. As long, however, as the primary wire remains in any one position, all electricity in the secondary wire disappears; but if in this position the strength of the primary current should be increased or diminished, momentary currents in the secondary wire would again mark the changes in the primary, the increase causing an inverse, and the decrease a direct current. Hence we conclude that *a current which begins, a current which approaches, or a current which increases in strength, induces an inverse momentary current in a neighboring circuit;* and that *a current which stops, a current which retires, or a current which decreases in strength, induces a direct momentary current in a neighboring circuit.*

In experiments like the above it is better to wind the primary and secondary wires on bobbins, so as to form coils, as shown in Figure 105. The primary coil *P* is made of coarse wire, and the secondary coil *S* of fine wire, as in the induction coils already described (274, 275). If the primary coil be placed in the circuit of a cell or battery, and if the secondary coil be connected with a galvanometer, a momentary inverse current appears in *S* when *P* is put inside it, and a momentary direct one when it is taken out; or if, while *P* remains in *S*, the strength of the primary current be altered, the galvanometer indicates the induction of currents according to the principles already stated.

Fig. 105.

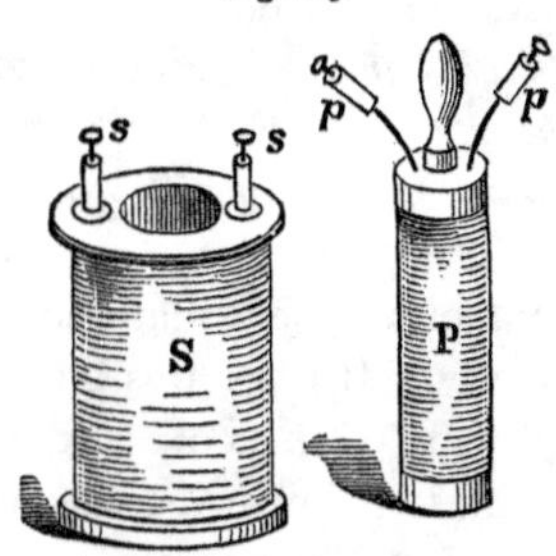

Current induction is probably only a phase of magnetic induction, since we have seen (237) that any wire through which a current is passing is slightly magnetic, and that such wires are powerfully magnetic when wound in coils. We have seen also (271) that, when a magnet is brought near to or removed from a conducting wire, it excites an inverse or a direct current in the wire; and that when soft iron placed within a helix is gaining or losing magnetism, momentary currents, inverse or direct, are excited in the wire.

279. *Extra Current.*—Not only does a galvanic current induce electricity in a neighboring current, but it also acts inductively on itself. When contact is broken in a battery circuit a spark is seen. When the wire is short the spark is feeble, but it grows brighter with the length of the circuit, especially when the wire is wound in a coil. The current is not stronger in the latter case, but weaker, as the galvanometer will show; and the real cause of the brighter spark

with the longer circuit is to be found in the induction of the primary current on the various parts of the circuit, exciting *extra currents*, as they are called, in the primary wire. Experiments have shown that *at the instant a galvanic current begins and ends, extra currents are induced by the action of the several parts of its circuit upon each other, that at the beginning of the current being inverse, and that at the end direct.* The effect of the extra current on the direct induced current of the secondary coil is to lessen its tension very decidedly. It does this by prolonging the cessation of the magnetism of the core and of the current in the primary coil, and thus impairing the suddenness of this change.

280. *The Magnetism of Rotation.* — It was long ago observed that, when a magnetic needle was made to oscillate above a copper plate, it came to rest sooner than it did otherwise. The oscillations were made in the same time as when away from the plate, but they were less in extent; the plate seeming to act as a damper to the motions of the needle. Arago reasoned from this that the needle at rest would be influenced by the plate in motion, and experiment confirmed the surmise. He made a copper disc revolve with great rapidity under a needle, the middle of the needle being directly above the centre of the disc. As expected, the needle was deflected in the direction of the motion of the disc; and the deflection increased with the rapidity of the motion, until at length the needle turned round after the disc. This action of the revolving disc was called the *magnetism of rotation*, and the name has been since retained. It was first explained by Faraday, who proved it to arise from the reaction of currents induced in the disc by the magnet. As the copper plate is a continuous conductor, currents will be started in it by its rotation under the magnet (271). These currents are found to flow in the neighborhood of the needle in such a direction as to deflect it in the direction in which the plate is moving.

SUMMARY.

Electricity can be developed by magnetism, either by moving a conductor near a constant magnet, or the magnet near the conductor; or by changing the strength of the magnetism in a magnet which is near a conductor.

Electricity developed by magnetism is called *magneto-electricity*, and an instrument for developing it is called a *magneto-electric machine*. In all ordinary machines of this kind the electricity is induced by developing and destroying magnetism in soft iron placed inside a helix. This may be effected by using a permanent magnet, or by means of the electric current. (271, 272.)

When the magnetism is developed and destroyed by means of the current, the instrument is usually called an *induction coil*. The most important machine of this kind is the *inductorium*, or Ruhmkorff's induction coil. It consists of a bundle of iron wires surrounded by a helix of thick copper wire, through which the primary current is sent in a succession of rapid impulses. This primary coil is surrounded by a much longer coil of very fine wire, in which the secondary current is induced. (275.)

The most important magneto-electric machine of the first class is the one recently invented by Wilde. In this machine the current developed by rotating an armature between the poles of a series of permanent magnets is made to develop much more powerful magnetism in an electro-magnet, which in turn is made to develop a current by means of a second rotating armature. In this way a magnet indefinitely weak may be made to develop a current of indefinite strength. (273.)

Magneto-electricity has much greater intensity than voltaic electricity.

Magneto-electricity may also be developed by means of

the current by alternately carrying continuous conductors near to and away from a wire through which a current is passing; or by leaving the conductor stationary near the wire, and alternately stopping and starting a current in the wire. This is called *current induction.* It is probably only magnetic induction, since the wire through which a current is passing is always magnetic. (278.)

The *extra current* which appears on stopping or starting a current in a wire is due to the inductive action of one part of the circuit upon another. (279.)

The movements of a magnetic needle when suspended over a rotating copper disc are explained by the action of the currents induced in the disc by its rotation under the magnet. (280.)

THERMO-ELECTRICITY.

281. *Electricity developed by Heat.* — We have seen the power of the current to develop heat (261), and we shall now see that heat has power to develop an electric current. The electricity so developed is called *thermo-electricity* (*heat* electricity).

Take a copper wire, cut it in two, and fix each half in one of the binding-screws of a galvanometer. Heat one of the free ends to redness, and press it against the other, and a current will be generated, passing at the junction from the hot to the cold end, as shown by the needle. Perform the same experiment with two pieces of platinum wire, and the current is stronger. In almost all cases where portions of the same metal at different temperatures are pressed together, a current is produced.

Currents also appear when two portions of the same metal or piece of metal have different structures, and the point where both structures meet is heated. If, for instance, one piece of wire be hard-drawn and the other part annealed (17), a current is produced when the point of separation between the hard and soft part is heated. The same takes place if one part of the wire be hammered or twisted, and the other part not.

Fig. 106.

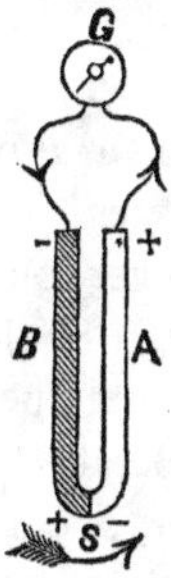

When the point of junction of any *two* metals is heated, a current is always produced. When a bar of antimony, *A*, is soldered to a bar of bismuth, *B*, (see Figure 106,) and their free ends are connected with a galvanometer, *G*, a current passes from the bismuth to the antimony when the junction is heated. When *S* is cooled by applying ice, or otherwise, a current in the opposite direction is produced. Such a combination of metals is called a *thermo-electric pair.* Metals like antimony and bismuth, which have a crystalline structure, are best suited for a thermo-electric pair.

282. *Thermo-electric Battery.* — One bismuth-antimony pair has very little power. To obtain a stronger current several pairs are united, as shown in Figure 107. The heat in this case must be applied only to one row of soldered faces. The strength of the current depends on the difference of temperature of the two sides; and to increase it to the maximum the one series must be kept in ice or in a freezing mixture, whilst the other is exposed to an intense heat. As in the galvanic battery, the electric force is proportionate to the number of pairs. At best, however, it is small, and the galvanometer used to measure it must be a very delicate one.

Fig. 107.

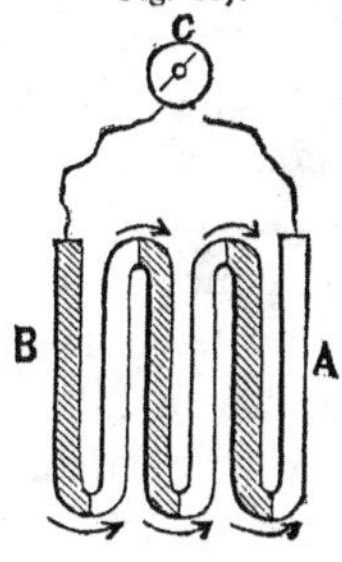

When a great number of pairs are formed into a battery, they are usually arranged as shown in Figure 108, which

Fig. 108.

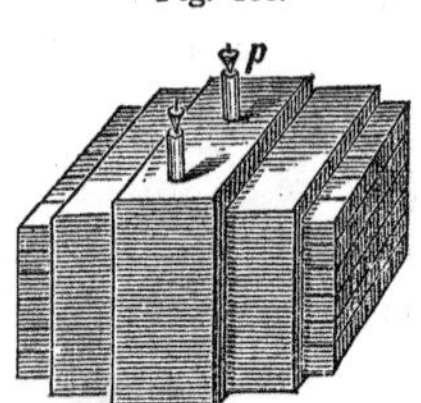

shows one of thirty pairs. The odd faces, 1, 3, 5, etc., are exposed on one side, and the even faces, 2, 4, 6, etc., on the other. The terminal bars are connected with the binding-screws. The interstices of the bars are filled with gypsum to keep them separate, and the whole is put in a frame of non-conducting material.

Such a battery, in connection with a sensitive galvanometer, forms a most delicate thermometer; showing, however, only differences of temperature between the two faces. So long as the opposite faces are exposed to the same temperature, no current is produced; but if the temperature of one side becomes higher than that of the other, a current is at once indicated. If the hand, for instance, be brought near one side, the needle shows a current; or if a piece of ice be held near, a current is also shown, but moving in the opposite direction.

SUMMARY.

Take pieces of two metals, and connect one end of each by a wire, and bring the other ends together. On heating the point of contact, a current is developed. Electricity thus generated is called *thermo-electricity*, and the metals thus connected constitute a *thermo-electric pair*. Such a pair can be formed of two pieces of one metal, provided these are in different conditions. Antimony and bismuth form the best thermo-electric pair. (281.)

Several such pairs connected form a *thermo-electric pile*, or battery. Such a pile is a very sensitive thermometer, since a current is developed by the slightest difference of temperature between the two faces. (282.)

FRICTIONAL ELECTRICITY.

283. *Electricity developed by Friction.* — When a cat's back is stroked on a cold, dry day, in a darkened room, sparks are obtained which at once indicate the development of electricity. If a well-dried rod of glass or gutta-percha be rubbed with a piece of silk or flannel, similar sparks appear. Hence electricity may be developed by friction. Such electricity is called *frictional electricity.* It is found by experiment that, when any two dissimilar bodies are rubbed together, electricity is developed; but when the substances are conductors of electricity, the force thus developed passes off silently through the hands and body. In order to detect it, the substances rubbed together must be held by insulating handles, that is, handles which do not conduct electricity.

284. *The Electrical Machine.* — In studying frictional electricity it is desirable to have an apparatus suitable for generating it. Such an apparatus is called an *electrical machine.* One of the best forms of it is shown in Figure 109. It consists of a thick plate of glass, insulated from

Fig. 109.

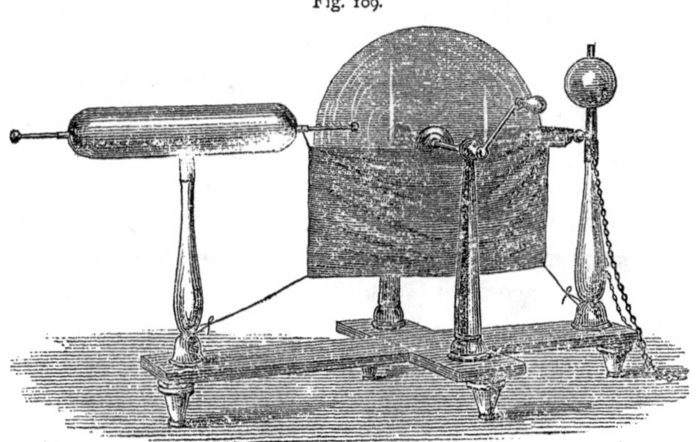

the floor, and turned by a crank. At one end there is a glass standard surmounted by a brass ball. From this standard project two brass strips in the form of a clamp, which hold the rubbers against the glass plate. These rubbers are pieces of wash-leather or woollen cloth, covered with an amalgam of mercury, lead, and tin. At the opposite end, on a glass support, is a long cylinder of brass with rounded ends. This cylinder is the *prime* or *positive* conductor. The brass ball connected with the rubber is the *negative* conductor. It is necessary that the plate and conductors of the machine be well insulated. In dry and frosty weather glass insulates very well. At all other times it becomes covered with a very thin, scarcely visible layer of moisture, which very much impairs its insulating power. In order to insure dryness it is necessary to rub the standards with a warm flannel before using the machine. The deposition of moisture is greatly lessened by coating the glass with shel-lac.

285. *Quantity and Intensity of Frictional Electricity.*— With a medium-sized electrical machine of this kind, sparks are readily obtained two inches long by presenting a conducting substance to the ball of the prime conductor. Machines have been constructed powerful enough to give a spark two feet in length. We thus see that frictional electricity has great tension (225). Its quantity, on the other hand, is next to nothing. This is shown by connecting the positive conductor with one end of the wire of a moderately delicate galvanometer, and the negative conductor with the other end, and working the machine. The needle will be deflected only one or two degrees. The great intensity and the small quantity of frictional electricity place it in striking contrast with voltaic electricity.

If a galvanic pair consisting of an iron or copper wire about $\frac{1}{18}$ of an inch in diameter, immersed about an inch in a little water to which has been added one drop of sulphuric acid, be connected with the above galvanometer, it will cause the needle to move six or eight degrees; but the electricity has so little tension that it cannot effect a discharge through the air even at a microscopic distance. Faraday has calculated that a wire of platinum and one of zinc, $\frac{1}{18}$ of an inch thick, immersed $\frac{5}{8}$ of an inch in four ounces of water acidulated with one drop of sulphuric acid, will produce as great a quantity of electricity in three seconds as thirty turns of a 50-inch plate machine.

The positive conductor of an electrical machine answers to the positive pole of a galvanic battery, and the negative conductor to the negative pole, and the friction on the plates to the chemical action in the cells. With the galvanic battery an enormous quantity of electricity is obtained of slight tension; with the electrical machine, a small quantity of enormous tension.

286. *The Electroscope.* — If a pith ball hung by a silk thread from a glass rod be brought near the ball of a prime conductor, it is at first briskly attracted and then as briskly repelled. This power of attracting light bodies is one of the most striking features of frictional electricity, and deserves especial study. It grows out of its high tension. The electricity at the unconnected poles of a powerful galvanic battery can be detected only by the most delicate apparatus, yet it can be shown to exist. The power of the electricity developed by friction to attract and repel light bodies furnishes the most ready means of detecting the presence of this electricity, as the needle furnishes the most ready means of detecting the presence of voltaic electricity.

An instrument constructed on this principle for the detection of frictional electricity is called an *electroscope.*

Figure 110 represents a common and convenient form of electroscope. It consists of a brass conducting-rod supporting a graduated semicircle, in the centre of which is a movable index made of very light wood, with a pith ball at the end. When it is attached to the prime conductor of the electrical machine, the pith ball is repelled as soon as the plate is turned.

Fig. 110.

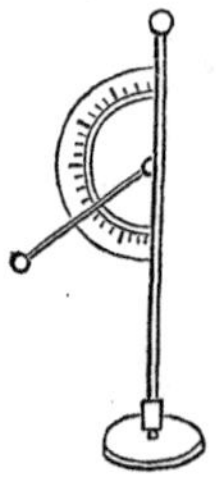

Figure 111 shows a more delicate electroscope. It consists of a hollow glass ball, the neck of which is covered by a brass cap. Through this cap, but insulated from it, passes a brass rod having a brass ball at its upper end and two narrow strips of gold-leaf suspended from its lower end. If the brass ball is brought near a body charged with electricity, the strips of gold-leaf repel each other, as in the figure.

Fig. 111.

As the gold-leaves are very easily torn, care must be taken not to communicate to them too strong a charge. In using this electroscope an instrument called a *proof plane* (see Figure 112) is often convenient. It consists of a small disc of gilt paper insulated by a glass rod. It is used by bringing it in contact with the electrified body, and then with the brass ball of the electroscope. The gold-leaves will immediately diverge, and, as only a small charge can thus be communicated, there is no danger of injuring them.

Fig. 112.

287. *The Electric Forces on the Positive and Negative Conductors act in Opposite Directions.* — Insulate both conductors of the machine, and charge them with electricity by turning the plate. Bring a pith ball suspended by a silk thread in contact with the positive conductor, and it is soon repelled. Take it now to the negative conductor, and it is strongly attracted. Discharge now the pith ball by taking it in the hand, and again bring it in contact with the negative conductor, and it is repelled;

but on taking it to the positive conductor it is attracted. We see then that a ball which is repelled by the force on one conductor is attracted by the force on the other. In other words, the forces on the two conductors act in opposite directions.

These opposite electrical forces are called *positive* and *negative* forces.

288. *Both Electrical Forces are always developed together.* — It is found to be impossible to develop one of these forces without at the same time developing both. The positive force always appears upon one of the substances rubbed together, and the negative force always appears upon the other. The force that acts in the same way as that upon the prime conductor of an ordinary electrical machine is called positive electricity, and the opposite force is called negative electricity. Of course, in order that both the forces should be detected, both of the substances rubbed together must be insulated.

The force that appears upon each of the substances depends upon their nature. When any substance, as glass, is rubbed with different substances, the same force does not always appear upon it.

289. *Induction.* — If an insulated copper ball is connected with the prime conductor of the machine, and a small insulated conductor is placed near it (see Figure 113), on developing electricity and examining the condition of the insulated conductor, opposite electrical forces will be found to be developed upon its ends. On the end next the ball, negative force will be found; on the end farthest from the ball, positive force.

Fig. 113.

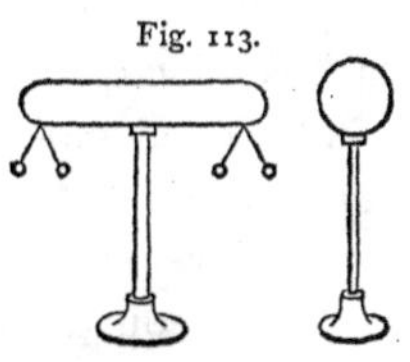

This action of a charged body upon a body near it is called *induction.* The insulated conductor is said to be *polarized.*

A charged body polarizes all insulated conductors near it; that is, it develops upon them opposite electrical forces at opposite points. It always develops on the part of the conductor nearest it a force opposite to that with which it is itself charged; and on the part of the conductor farthest from it, a force the same as that with which it is charged.

If one copper ball is connected with the positive conductor and another with the negative conductor, and a row of insulated conductors is arranged between these balls, as

Fig. 114.

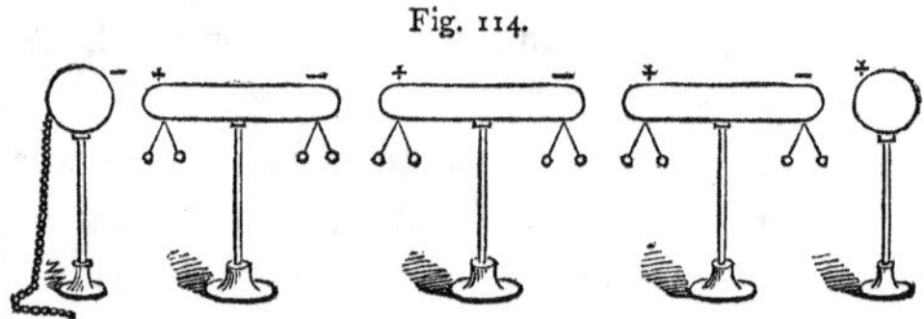

shown in Figure 114, on developing the electric force each of these insulated conductors will become polarized. The ends of the conductors towards the ball connected with the positive conductor are all negative, and the opposite ends positive.

Place three insulated conductors end to end between the two copper balls. First arrange the conductors quite near each other, and develop electricity. Sparks will pass between the ends of the conductors. Completely discharge each conductor by passing the hand over it, and place the middle conductor quite near the conductor on one side, and at some distance from the conductor on the other side. Again develop electricity, and a spark passes between the middle conductor and the conductor nearest to it. Let the condition of the conductors between which the spark has passed be now examined by means of a light pith ball or a proof plane and electroscope. Previous to the passage of the spark there were two forces on each conductor; but after the passage of the spark only one force is found on each conductor. The force on each conductor is the same as that first developed on the ends farthest from each other; and of course those forces are of opposite kinds. The opposite forces, then, upon the ends of the conductors facing each other, have passed from the conductors and become neutralized. Let now the conductors on each side be brought up so as almost to touch the ends of the middle conductor, and the balls so placed as almost to touch the ends of the outside conductor. On turning the machine sparks pass between the conductors, and on examining them scarcely any force is found upon them. If the conductors and balls are brought quite in contact, no force is found upon them when the machine is turned.

We see, then, that when insulated conductors are placed end to end near a charged body, opposite forces are developed at the opposite ends of each conductor; and that, when two conductors are so arranged that the spark can pass between them at one end and not at the other, only one force is found on each, and that one the force which was first developed on the ends between which the spark did not pass; also that, when several conductors are so arranged between the balls that a spark can readily pass from each end, no force remains on each.

When the two opposite forces exist on a conductor, it is said, as we have already learned, to be *polarized;* when only one force exists on it, to be *charged;* and when no force exists on it, to be *neutral.* When a force which has been developed on an insulated conductor passes off, it is said to be *discharged.*

When an insulated conductor is brought near a charged body, it is first *polarized,* and the nearer it is brought, the higher the polarization rises. If the conductor is so situated that it can discharge its force at only one end, it becomes charged with the same electric force as the body originally charged; if it discharges from the opposite end, it becomes charged with the force opposite to that on the originally charged body. If the conductor is so situated that it can discharge quite readily at both ends, but more readily at one end than at the other, there will be three steps in the process. The conductor will first become polarized, then charged, and finally neutralized.

If the conductor is so situated that it can discharge quite readily, and with equal readiness, at each end, there will be only two steps in the process. It will be first polarized, and then neutralized.

290. *The Polarization of the Insulated Conductor depends on the non-conducting Medium which separates it from the Charged Body.* — Charge a metallic disc, and bring it near the ball of the gold-leaf electroscope ; the leaves diverge, owing to the electricity developed on them by induction. Introduce a thick cake of shel-lac between the disc and the ball, and the leaves diverge still more, showing that the polarization has risen higher. Faraday has shown that the polarization of a body changes whenever a different non-conductor occupies the space between it and the charged body. We therefore conclude that the polarized condition of a body depends upon the non-conducting medium which separates it from the charged body.

291. *Faraday's Theory of Induction.* — Faraday supposes that each molecule of a substance acts like an insulated conductor, and that the main difference between a conductor and a non-conductor is, that the molecules of the former are so arranged that they can readily discharge their forces into one another, while the molecules of the latter discharge their forces into one another with difficulty. It is supposed also that the molecules of conductors are polarized more easily than those of non-conductors.

The electric force, then, like cohesion and adhesion, must be regarded as one of the molecular forces. Its first action is to polarize all the molecules in the neighborhood; that is, it develops opposite electric forces on opposite parts of each molecule. The molecules of conducting substances are polarized readily, and those of non-conductors with difficulty. The molecules, on becoming polarized, tend to discharge their forces into one another at opposite points. The molecules of conductors discharge their forces easily, and those of non-conductors with difficulty. When the molecules are entirely surrounded by molecules of the same kind, discharge takes place with equal readiness at any point, and the neutral state follows that of polarization. When the molecules are surrounded on one side by conducting molecules, and on the other side by non-conducting molecules, discharge takes place more readily on one side than on the other, and polarization is always followed by the charged state.

Let us now examine the condition of the balls and the conductors in one of the preceding experiments. It will be remembered that one of the balls was connected with the positive conductor, and the other with the negative conductor, and that the insulated conductors were separated by a stratum of air. On turning the machine, electricity is developed, and all the molecules along the line of the conductors, of the insulated conductors, and of the air between them, are first polarized. The molecules of the conductors immediately discharge their forces and become *neutral;* the molecules of the non-conducting air do not discharge their force so readily, and still remain *polarized;* while the molecules on the faces of the balls and of the insulated conductors discharge their forces readily into the conducting molecules on one side of them, and with difficulty into the non-conducting molecules on the other side, and are therefore *charged.* It will be remembered that we have already found by experiment that these faces of the conductors are charged.

As more electric force is developed, the charges on the molecules at the faces of the conductors increase, and the polarization of the molecules of the air also increases. This polarization finally rises so high that these molecules also discharge their forces into one another, and the whole line becomes neutral. The discharge of the non-conducting molecules of the air into one another gives rise to the spark.

292. *Disruptive and Conductive Discharge.*—Place a conductor near the prime conductor of the machine, and hold a ball between the two by means of a silk thread. This little insulated conductor will be of course first polarized. It then flies to the prime conductor and discharges its force on that side, then back to the other conductor and discharges its force on the opposite side. On thus becoming neutralized, it is again immediately polarized, and repeats the movements already described. We thus see that a light insulated conductor, when free to move, is thrown into commotion upon discharging its forces. If then the molecules of bodies act like insulated conductors, it is reasonable to expect that they will also be thrown into commotion on discharging their forces into one another. We have already noticed the fact that fine wires are heated on the passage of the electric current through them. We shall soon learn that heat is the result of the commotion of molecules of matter. We must, therefore, conclude that the molecules of conductors are thrown into commotion on discharging their forces. The sound, light, and heat of the spark indicate that the molecules of non-conductors are thrown into still greater commotion on discharging their forces.

Because of this greater commotion of the molecules, the discharge which takes place between the molecules of non-conductors is called *disruptive discharge.* The discharge which takes place between the molecules of conductors is called *conductive discharge.*

293. *The Electric Bells.* — The experiment with the electric bells gives a very pretty illustration of disruptive discharge. Three bells are hung from a brass rod; the two outside ones by means of brass rods, and the middle one by a silk cord. Two brass balls are hung between the bells by silk cords. The middle bell is connected by a chain with the floor, and thus to the negative conductor. The rod is then connected with the prime conductor, and electricity is developed. The outside bells become charged, and the balls polarized. The balls in discharging their forces are thrown into commotion, and are dashed against the bells on each side, giving a musical sound instead of the sharp snap of the spark, which is undoubtedly caused by the dashing together of the molecules of the air.

294. *There are no Non-conductors.* — From what has been stated it follows that there is no such thing as a non-conductor of electricity. A non-conductor is only a very poor conductor. A better division of substances would be into good and bad conductors of electricity.

This may be shown also by the following simple experiment. Connect the inner and outer coatings of a charged Leyden jar (302) by a long wire, which near the coatings is bent towards itself within a fourth of an inch (see Figure 115); and the greater part of the discharge, instead of passing through the long wire, the course of which is cut off by a dotted line in the figure, leaps across at the point E; the proportion being greater, the nearer the wire is at the bend, and the longer the course of the wire.

Fig. 115.

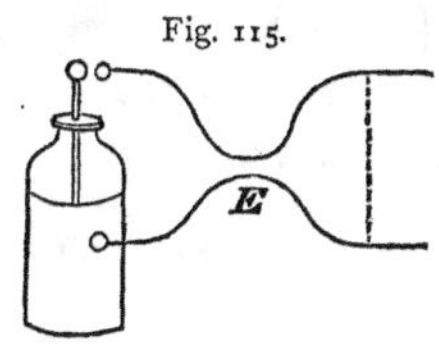

295. *The Electric Current explained by these Principles.* — In the voltaic battery the electric forces are generated by chemical action (267). When the poles of the battery are connected by a copper wire, the molecules of the wire become polarized and then discharge their forces, again

12

become polarized and discharge, and so on. The electric current, then, is only the successive polarization and discharge of the molecules of the wire. The polarization evidently starts from each pole of the battery and runs along the wire to its centre, and is then followed in the same way by discharge.

296. *Electrolytic Discharge.* — We have seen that, when the electric current is sent through a compound liquid, the latter is decomposed. According to our theory, the molecules of the liquid first become polarized and then discharge into one another. It is then evident that the molecules of compound liquids, on discharging into one another, are not only thrown into commotion, but actually broken up.

As fast, however, as they are broken up, except at the end of the line, they are re-formed. We will suppose that water is the electrolyte. Each molecule of water is composed of two atoms of hydrogen and one of oxygen (52). On becoming polarized the molecules of water seem to face round so that the hydrogen atoms are turned towards the negative electrode, and the oxygen atoms towards the positive electrode. Representing the positive electrode by + and the negative by —, the condition of a row of molecules of water between the electrodes, on becoming polarized, may be figured thus: —

$$+\,\Theta^{\mathrm{H}}_{\mathrm{H}}\quad \Theta^{\mathrm{H}}_{\mathrm{H}}\quad \Theta^{\mathrm{H}}_{\mathrm{H}}\quad \Theta^{\mathrm{H}}_{\mathrm{H}}\quad \Theta^{\mathrm{H}}_{\mathrm{H}}\,-$$

On discharging their forces the oxygen and hydrogen atoms separate, but recombine everywhere, except at the ends of the line.

The following may represent the arrangement of the molecules after they have discharged the forces developed by polarization: —

$$+\,\Theta\quad {}^{\mathrm{H}}_{\mathrm{H}}\Theta\quad {}^{\mathrm{H}}_{\mathrm{H}}\Theta\quad {}^{\mathrm{H}}_{\mathrm{H}}\Theta\quad {}^{\mathrm{H}}_{\mathrm{H}}\Theta\quad {}^{\mathrm{H}}_{\mathrm{H}}\,-$$

We thus see that, while decomposition has taken place along the whole line, the oxygen appears only at the anode, and the hydrogen at the cathode.

The following symbols may represent the arrangement of the molecules of sulphate of copper after polarization and after discharge:—

$$+\ SO_4Cu\quad SO_4Cu\quad SO_4Cu\quad SO_4Cu\ -$$
$$+\ SO_4\quad CuSO_4\quad CuSO_4\quad CuSO_4\quad Cu\ -$$

The discharge which takes place between the molecules of a compound liquid is called *electrolytic discharge.* It always results in the breaking up of the molecules.

297. *The Charge on a solid Insulated Conductor is always on the Surface.*—To an insulated copper ball are carefully fitted two hemispherical metallic caps provided with insulating handles. The caps are placed upon the ball, and the whole apparatus is charged. The caps are then removed and examined, and are found to be charged, while not the slightest trace of a charge is found on the ball inside.

According to the theory that the molecules of all substances act like insulated conductors, we see that the charge must reside on the surface. Since the molecules in the interior of a conductor discharge with equal readiness on every side, there is no chance for them to become charged; but at the surface the molecules have conducting molecules on one side of them and non-conducting molecules on the other, and hence their polarization must be followed by the charged state.

298. *Distribution of the Charge on the Surface.*—It is found by experiment that, when a spherical conductor is charged and placed in the centre of a room, the charge is distributed uniformly over its surface; and that, when an oblong conductor is charged and placed in a similar situation, the charge is found to accumulate at the ends.

Again, when an insulated spherical conductor is charged and a conductor brought up to one side of it, the charge is found to accumulate on that side. The electric force, then, tends to accumulate where there is least resistance to its polarization. For it is evident that there is less resistance to its polarization in the direction of the conductor than in any other direction, since molecules of conductors are charged more easily than those of non-conductors.

299. *Polarization rises highest in the Direction of the least Resistance.* — When a conductor is brought near enough to a charged body, a spark passes. The resistance to polarization is less in this direction than in any other, and the fact that the spark passes in this direction shows that the polarization has risen higher than elsewhere; otherwise the molecules would not have discharged their forces at this point.

300. *Why the Charge accumulates at the Ends of Oblong Conductors.* — We have already seen why, according to our theory, the charge must reside on the surface of a body, and the fact that we have just established — that polarization rises highest in the direction of the least resistance — explains why the charge accumulates at the ends of oblong conductors. When any part of the conductor becomes charged, the polarization rises highest in the direction of the least resistance, that is, towards the ends of the conductor. Of course, the higher the polarization the higher the charge which follows it. Hence the charge is higher at the ends than at the sides of an oblong conductor. In the case of the sphere, any part of the surface is just as much an end as any other part. Hence the charge is distributed uniformly over the surface of a sphere, when it is so situated that it can carry on polarization with equal readiness in every direction.

301. *The Charge which a Conductor can receive depends upon its Facilities for carrying on Polarization.* — This fact

is illustrated by the following piece of apparatus. It consists of three cups made to fit closely within one another. The outer and inner ones are of tin; the middle one, which is higher than the others, is of glass. In the centre of the inside tin cup there is an upright glass tube, within which is a brass chain attached to the bottom of the cup, and having a brass ball at the other end. Remove the outside tin cup, and charge the inside tin cup by bringing the brass ball near the prime conductor of the machine. The glass cup should be placed on a glass stand. But few sparks will pass between the ball and the prime conductor, showing that the cup receives but a small charge. Discharge the cup, replace the outside tin cup, connect the latter with one of the conductors of the machine, and bring the ball of the inside cup near the other conductor of the machine. On developing the electric force a large number of sparks pass between the ball and the conductor, showing that the cup has received a much larger charge in the second case than in the first.

In the first case, the charge on the cup must polarize the molecules of the glass cup and of the air about it; in the second case, those of the metallic cup and the conductors outside. Since the molecules of the metallic cup and the conductors are polarized much more easily than those of the air, it is evident that polarization can be carried on by the cup much more easily in the second case than in the first.

Hence the greater facilities an insulated conductor has for carrying on polarization, the greater the charge it can receive.

302. *The Leyden Jar.* — Replace the two metallic cups by tinfoil, and the apparatus just described becomes an ordinary Leyden jar. This jar is charged by connecting its outer coating with one conductor of an electrical machine, and the inner coating with the other, and developing electricity.

303. *The Condition of the Jar when Charged.*—When the jar is arranged as above described, and electricity is developed, the molecules all along the line of the conductors, of the tinfoil coatings, and of the glass between them, are first polarized. Then they are all neutralized, except the layers on the inside and outside coating, which are evidently charged, and those of the glass, which are still polarized. As the molecules next to the glass on the inside discharge from the opposite part from those next the glass on the outside, the two layers of molecules are evidently charged with opposite electric forces. The layer of molecules of either coating nearest the glass is charged with the same electric force as that of the conductor with which it is connected. If the charge rises high enough, the molecules of the glass become so strongly polarized that a disruptive discharge takes place between them.

When the jar is disconnected from the conductors of the machine, the charge still remains on the surface of the coatings next the glass, since polarization can be carried on more readily through the glass than through the air. The jar can be safely handled, if care is taken not to touch both coatings with any part of the body. If both are thus touched, a violent shock is received.

The jar may be discharged by means of the *discharger*, which consists of two bent brass arms connected by a movable joint and having brass balls at their ends. It is fastened at the joint to a glass handle. To discharge the jar, hold the discharger by the glass handle, and bring one ball in contact with the outer coating and the other ball near the knob connected with the inner coating. As soon as polarization can be carried on more readily through the air which separates the balls than through the glass, the charge leaves the surface of the coatings next the glass and collects upon the balls. Disruptive discharge then takes place between the balls.

304. *The Leyden Battery.*—The amount of charge which a Leyden jar can receive, other things being equal, evidently increases with the size of the coatings. The area of the coatings can evidently be increased, either by making the jar larger, or by connecting together several smaller jars. The latter arrangement constitutes a *Leyden battery.* Like the cells of the voltaic battery, the jars can be connected in two ways: (1.) the outer coating of one may be connected with the inner coating of the next, and so on throughout the series; or (2.) the outer coatings may all be connected together, and also the inner coatings. In the first case, the battery is discharged by bringing the inner coating of the first jar in contact with the outer coating of the last; in the second case, by bringing the connected outer coatings in contact with the connected inner coatings. Like the voltaic battery, when the Leyden battery is arranged in the first way, it gives electricity of the greatest intensity; and in the second way, electricity of the greatest quantity.

The spark obtained from a powerful Leyden battery can be made to imitate on a small scale all the effects of lightning. It can be made to split tough bits of wood, shiver glass, and the like.

If a body is so situated that it cannot carry on polarization, it can receive no charge. This accounts for the fact that hollow spherical and cylindrical conductors become charged only on their outer surfaces.

305. *The Effect of Points on a Conductor.*—It is found to be impossible to charge a conductor when a sharp point projects from it, or is held near it. The point conveys away the electric force silently. If the hand is held in front of the point when the electricity is developed, a current of air is distinctly felt setting off from the point. If a lighted taper is held near the point, the flame is blown away from it. The electric force is then evidently carried

off by the molecules of the air which form the current. The silent discharge effected by points is evidently a kind of disruptive discharge; and, since the force is conveyed away by the movement of the molecules of the air, it is called *convective discharge.* Since in a darkened room a star of light is seen upon a point held near a powerful electrical machine while in action, this silent discharge is also called *glow discharge.*

The charge rises so high at the point that the molecules of air just about it are strongly polarized. They then seem to act like little pith balls. The molecules directly in front of the point are first attracted and then repelled; while those just behind are in turn drawn to the point and then driven from it, giving rise to a current of air from the point.

306. *The Electric Wheel.*—As each molecule is repelled from the point, it also repels the point itself, which, if free to move, ought to move as well as the molecules of air. This explains the action of the *electric wheel,* which consists of a number of points all bent round in the same direction, as shown in Figure 116. The wheel is poised so as to turn easily, and when connected with the prime conductor of the machine in action, it rotates rapidly, each point moving backward.

Fig. 116.

307. *Atmospheric Electricity.*—Franklin, with his famous kite experiment, first proved the identity of lightning and ordinary frictional electricity. It is well known that a cloud often becomes charged with electricity. How it becomes thus charged is not so well known. It seems probable that the friction of the cloud against the air has something to do with the development of the charge. When a cloud thus charged comes near a second cloud, the two act like the two coatings of a Leyden jar. When the charge rises high enough, the molecules of the air between the two

clouds become so highly polarized that disruptive discharge takes place, and gives rise to the flash of lightning. When we consider the extent of the coatings in this case, we are not surprised at the magnitude of the spark. When the clouds are so situated that we can see the whole length of the spark, it appears as *chain lightning*. When the discharge takes place behind a cloud, we see only the illumination of the cloud by the flash, and the appearance is called *sheet lightning*.

It often happens that a charged cloud and the earth act as the two coatings of an immense Leyden jar, and the disruptive discharge takes place between the cloud and the earth.

308. *The Use of Lightning-Rods.* — It has already been seen that polarization rises highest in the direction of the least resistance, and that discharge takes place when polarization rises highest. Hence, when a charge accumulates in a cloud above a house to such an extent that disruptive discharge takes place between the cloud and the earth, the lightning-rod offers a path of less resistance than the house; hence discharge will be more likely to take place through the rod than through the house. The rod must extend into moist earth, else some other part of the house may furnish an easier path to the earth than the rod. In this case, the discharge will leave the rod and take the path of least resistance. The dry soil of the earth is a poor conductor of electricity.

Trees are more often struck than the ground, because they are better conductors than the air, and hence offer less resistance to polarization.

Lightning-rods should always be pointed at their upper ends, that they may convey away the charge from the clouds silently, and as far as possible prevent its accumulation. The points of the leaves of trees and of the blades of grass undoubtedly convey an enormous amount of electricity silently from the clouds, and thus greatly diminish the liability of violent disruptive discharge.

309. *The Aurora.* — The aurora is undoubtedly an electric discharge high up in rarefied air. In what way the electricity in this case is developed we are as yet wholly ignorant.

The auroral arch surrounds the magnetic pole of the earth.

The discharge through rarefied air somewhat resembles the aurora, and it has been found that this discharge can be made to circle round the pole of a magnet, as the auroral arch does round the magnetic pole of the earth.

SUMMARY.

When unlike substances are rubbed together, electricity is developed. Electricity thus generated is called *frictional electricity;* and an instrument for generating it is called an *electrical machine.* The positive conductor of the machine corresponds to the positive pole of the cell, and the negative conductor to the negative pole; and the friction between the rubber and the plate takes the place of the chemical action.

Frictional electricity has slight quantity, but enormous tension; thus contrasting strikingly with voltaic electricity, which has slight tension, but enormous quantity. Magneto-electricity, in this respect, stands midway between the two. (283-285.)

Two opposite electrical forces are developed on working the electrical machine, one appearing on each of the conductors. It is impossible to develop one kind of electrical force without at the same time developing the opposite. (287, 288.)

The small quantity of frictional electricity renders a galvanometer unfit for examining it. It is best examined by its attraction and repulsion of light bodies, a property which is due to its intensity. An instrument for examining the quality of frictional electricity is called an *electroscope.* (286.)

A body is *polarized* when it has opposite electrical forces developed on opposite parts; it is *charged* when it has only one electrical force upon it; and it is *neutral* when it has none upon it.

A charged body polarizes all insulated conductors near it, developing upon the face of the conductor nearest itself the opposite electrical force to that with which it is itself charged. When several insulated conductors placed in a row are polarized by the same body, there will always be opposite electrical forces developed on the parts which face one another. When insulated conductors are thus polarized, they tend to discharge their opposite forces into each other, and thus to become neutral. If they are so arranged that they can discharge one of their forces more readily than the other, they become first charged and then neutral. If they can discharge both forces with equal readiness, they become neutral as soon as they have become sufficiently polarized to effect discharge. (289.)

The polarized condition of an insulated conductor depends upon the non-conducting medium which separates it from the charged body, as is shown by changing the medium. Every time the medium is changed, the polarized condition is changed. (290.)

According to Faraday's theory, in all electrical action the molecules of matter act like insulated conductors. When these molecules are so situated that they are readily polarized and discharged, the substance is called a *conductor;* when they become polarized and effect discharge with difficulty, a *non-conductor.* Thus non-conductors are only poor conductors. (291, 294.)

When a very light insulated conductor is suspended between two polarized conductors, it is first polarized, and then discharges alternately into each of the conductors. In effecting this discharge it is thrown into commotion. The light of the electric spark, and the heat developed in a conductor on the passage of the electric current, show that molecules are also thrown into commotion on discharging their forces. (292.)

When discharge takes place between the molecules of compound liquids, these molecules are not only thrown into commotion, but are also broken up, as is seen in *electrolysis.* (296.)

The discharge which takes place between the molecules of a non-conducting substance is called *disruptive discharge;* that between the molecules of a conductor, *conductive discharge;* and that between molecules of a compound liquid, *electrolytic discharge.* (292, 296.)

Polarization is carried on most strongly in the direction of the least resistance, as is shown by the passage of the spark between two conductors. The charge which a body can receive depends upon the readiness with which it can carry on polarization, as is shown in the case of the Leyden jar. (299, 301, 302.)

When a spherical conductor which can carry on polarization with equal readiness in every direction is charged, the electric force spreads uniformly over its surface; but when it can carry on polarization more readily in some directions than in others, the charge accumulates at those points. This explains why the charge of the Leyden jar remains on the surface of the coatings next the glass. Since polarization is carried on most readily in the direction of least resistance, the charge tends to accumulate at the ends of oblong conductors, and especially on points. (298, 300.)

The action of points on charged bodies is peculiar.

They convey the charge off silently by *convective discharge.* This explains one of the uses of lightning-rods, which are always pointed at the upper end. They thus convey off much of the charge from their neighborhood silently. They also furnish a path of less resistance between the cloud and the earth than the building. (305, 308.)

A cloud and the surface of the earth often act as the two coatings of a Leyden jar, and the air between as the glass of the jar. In this case the discharge takes place between the cloud and the earth. At other times two clouds act as the two coatings, and the discharge then takes place between the clouds. How the clouds become charged is not well known. (307.)

The *Aurora* seems to be an electric discharge in rarefied air. (309.)

CONCLUSION.

We have now become somewhat acquainted with the force called ELECTRICITY. We first found it in the wire that connected the carbon and zinc of a Bunsen's cell. We have learned that this force has two qualities which it does not always possess in the same proportions; the one enabling it to turn a magnetic needle, and called its *quantity;* the other enabling it to overcome resistance, and called its *intensity*.

A further study of the action of the electric current on the magnetic needle explained the operation of the galvanometer, which is the most delicate instrument for measuring the quantity of the electric force, and also of the needle telegraph. We next discovered that the electric current has power to develop magnetism in a piece of iron, and also in the wire through which it is flowing. A further examination of this subject explained the action of electro-magnets, and some of the ways in which they are used as a motive power, including Morse's telegraph, the electric fire-alarm, and electric clocks. A fuller consideration of electrolysis, which had already been illustrated in the decomposition of water and muriatic acid, explained the processes of electrotyping and electro-plating, and also the action of Bain's electro-chemical telegraph. We then noticed the power of the electric current to develop heat, and saw how this power can be employed in blasting and as a source of light.

Having thus considered somewhat the action of the electric force, we next sought after its sources. We found that whenever two unlike bodies are brought together, and any change is effected at their point of contact, electricity is generated. When this change is brought about by chemical action, we obtain voltaic electricity; by heat, thermo-electricity; by friction, frictional electricity. We found also that electricity may be developed by magnetism, and that it is then called magneto-electricity. While considering each of these kinds of electricity, the instruments employed for developing it were described, and their action explained.

While we were studying frictional electricity, we became acquainted with Faraday's theory of electrical induction, which explains all the phenomena of electricity by supposing that the molecules of matter act like little conductors more or less insulated according to the conducting power of the body.

This mysterious force, which we can generate on a small scale by the above methods, and which has become of the greatest importance in its practical applications, is developed on an enormous scale in nature by processes of which we know little or nothing. The sparks of our most powerful electrical machines, and the most brilliant discharge which we can obtain through a vacuum tube, are but miniature representations of the lightning and the aurora.

APPENDIX.

APPENDIX.

1. (page 17.) A nipper-tap is a convenient substitute for a stop-cock. By means of it a rubber tube can be pinched close. Two kinds are shown (full size) in Figures 117 and 118. In one the tube is compressed by a spring; in the other, by a screw.

Fig. 117.

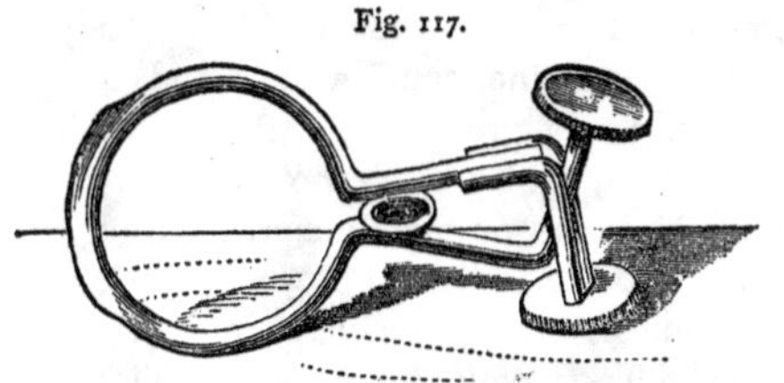

The U-tube used in this and similar experiments has near the bend a tubulature to which the rubber tubing can be fastened. A convenient size for U-tubes of this kind is about 8 inches for the length of the arms, and about $\frac{3}{8}$ of an inch inside diameter. The tube should be quite heavy, but not too thick.

Fig. 118.

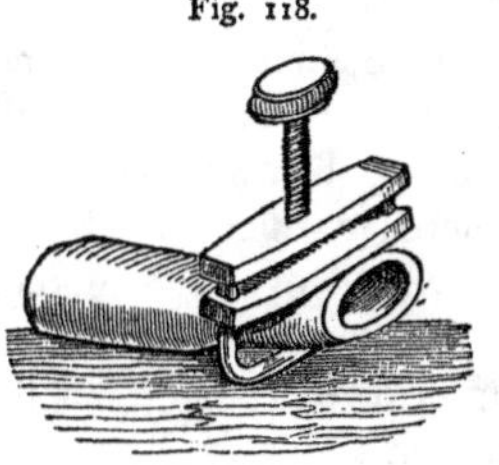

For some purposes it is convenient to have platinum wires passing into the tube near its closed end, for the attachment of battery wires. The wires should be separated about $\frac{1}{8}$ of an inch within the tube. These wires interfere with the use of the tube in none of the experiments. It is not necessary that the tube should be mounted on a stand; it can be held equally well in the clamp of an ordinary wooden retort-holder.

These U-tubes can be obtained at slight cost from any glass-blower. We have found that Mr. Charles Siefert, corner of Washington and State Streets, Boston, does such work very well, and at reasonable rates.

We would say generally that all chemical ware which need not be made to order can be obtained very promptly, and at fair prices, from George Quettier, 193 Greenwich Street, New York City. Wooden retort-holders and little adjustable stands, which are very convenient (see Figures 22, 27, etc.), can be best obtained at a cabinet-maker's. Mr. J. G. Bicknell, of Cambridgeport, does such work excellently and cheaply.

2. (page 29.) Gases, like ammonia and muriatic acid, which are absorbed by water, must be collected over mercury. For this purpose small glass cylinders are convenient, from 1 inch to 1½ inches in diameter, and from 5 to 7 inches high. They should be strong and should have the mouth well ground, and should be provided with covers of ground plate-glass. Such cylinders can be obtained from Quettier, of New York.

For the experiments of this book we think a mercury trough a needless luxury. An evaporating dish of Berlin porcelain answers every purpose. One five inches in diameter is large enough to use with the cylinders mentioned above. It should be set in a shallow dish of strong glass or earthen ware, to prevent loss of mercury in case of accident. It should then be filled nearly full of mercury. The cylinder to be used is next filled with mercury to the brim. The ground-glass cover is then placed on the mouth of the cylinder and held firmly with the right hand, the cylinder inverted, and its mouth plunged beneath the mercury in the evaporating dish. It is well to put the cylinder in a strong dish when you are filling and inverting it, to save any mercury which may be spilled. After the cylinder is filled and inverted over the mercury in the dish, the plate is slipped from its mouth, and the cylinder is securely fastened in the clamp of a retort-stand, so that its mouth is held about ½ or ¾ of an inch above the bottom of the dish.

To fill the cylinder with ammonia gas, a little aqua ammonia is put into a flask provided with a delivery tube, and gently boiled. To the delivery tube is attached a piece of rubber tubing to connect it with the cylinder. A short piece of glass tubing with its end slightly turned up must be fastened to the end of the rubber tube, so that the gas may be introduced into the cylinder through the mercury. On first boiling the aqua ammonia, a large amount of air passes over from the flask. The ammonia gas must not be collected until this has all passed over. To determine when this has taken place, put the glass tube at the end of the rubber tube into a vessel of water; when heat is first applied to the flask, bubbles of gas will rise from the end of the tube through the water as long as any air comes over. When the air is all over, the ammonia gas is absorbed by the water as fast as it comes over, and no bubbles rise to the surface. After the air ceases to come over, plunge the end of the glass tube into the mercury, and introduce it under the mouth of the cylinder. The gas will rise into the jar and displace the mercury.

A small bit of boxwood charcoal is now heated in the flame of a lamp, and plunged under the surface of the mercury a short time till it becomes cool, and then introduced into the jar.

Boxwood charcoal can be readily made by heating bits of boxwood in a copper flask till they cease to fume.

3. (page 33.) Gases, like hydrogen and carbonic acid, which are not absorbed by water, or but slightly so, can be collected by first filling the vessel with water, and inverting it with its mouth under water. The tube through which the gas is escaping is introduced under the mouth of the jar, and the gas rises and fills the jar.

For collecting gases that are not absorbed by water a *pneumatic trough* is useful and almost indispensable. This apparatus consists of a vessel deep enough to allow of filling and inverting under the water any of the jars ordinarily used. A shelf perforated with holes extends across the vessel about an inch under the surface of the water. After the jar is filled with water and inverted, it is set upon the shelf over one of the holes. The delivery tube is then passed into this hole, and the jar filled. It is convenient to have a shelf large enough to hold several jars after they have been filled with gas.

A pneumatic trough should be made of copper. Zinc is much cheaper, but less durable.

In the experiment described in section 44, the lower bottle should be first filled with carbonic acid, by first filling it with water and inverting it over water, and introducing into it the delivery tube of the bottle in which the gas is prepared (see Figure 38, page 103, and the accompanying description). When full, it should be closed with a good cork while its mouth is still under water, and placed in the position shown in Figure 15. The tube which is to connect the two bottles should then be passed through the cork of the bottle which is to be filled with hydrogen, and a cork accurately fitting the carbonic-acid bottle should be pressed upon the other end of the tube. The tube and bottle should then be filled with water, inverted over water, and filled with hydrogen by introducing into it the delivery tube of a hydrogen generator. When it is filled, close the tube with the thumb, and carry it to the bottle of carbonic acid. Remove the cork of this bottle, plunge the tube into it, and fasten it firmly by means of the cork already on the tube. The tube and bottles can then be arranged as in the figure, by means of a retort-holder.

Fig. 119.

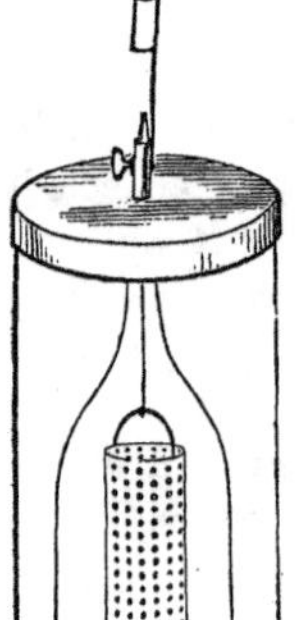

Hydrogen is most readily prepared by the action of dilute sulphuric acid on zinc. The most convenient apparatus for the purpose is the self-regulating *generator* shown in Figure 119. It consists of a glass vessel closed with a metallic cap. A bell-shaped glass vessel open at top and bottom is fastened to this cap by an air-tight joint. A tube closed by a stop-cock passes through the cap into the bell-shaped vessel. A copper bucket perforated with fine holes is hung from a hook inside this vessel. This bucket is filled with shreds of sheet zinc or bits of granulated zinc. The outer vessel is filled about two thirds full of dilute sulphuric acid (the ordinary oil of vitriol diluted with about ten parts of water). The metallic cap with the bell-glass attached is then put in its place, and the stop-cock opened. The air is first driven out, and the dilute acid coming in contact with the zinc begins to act upon it, and hydrogen is given off in abundance. When sufficient hydrogen has been obtained, the stop-cock is closed, the hydrogen generated collects in the upper part of the bell-glass, and drives the acid out. As soon as the acid is driven out, the action ceases until the stop-cock is opened again, and the hydrogen allowed to escape. By means of rubber tubing the hydrogen can be conveyed from the generator to the jar or vessel in which it is to be collected. When hydrogen is to be burned, the greatest care must be taken that all the air is driven out of the apparatus before the gas is collected. Hydrogen may also be prepared from zinc and sulphuric acid in a bottle similar to the one used for the preparation of carbonic acid.

The action of sulphuric acid on zinc is shown by the following equation: —

$$Zn + H_2 SO_4 = Zn SO_4 + 2H.$$

When sulphuric acid and water are mixed, great heat is developed. Care must be taken, therefore, to mix them very gradually. It is better to pour the acid into the water, than the water into the acid.

4. (page 33.) Lime-water is prepared by carefully slaking ordinary lime, then dissolving it in water, and filtering it. Water will dissolve but a small quantity of lime.

When lime-water is mixed with carbonic acid, the reaction is as follows: —

$$CaO, H_2O + CO_2 = CaO, CO_2 + H_2O.$$

The white carbonate of calcium formed is insoluble in water; hence the liquid becomes milky.

5. (page 34.) The bell-jar can be filled with hydrogen by the method described above (3). A ground-glass plate is held to its mouth when it is removed from the trough. When brought over the cup, the plate is removed, and the jar lowered into the position shown in the figure. The hydrogen is so light that it has little tendency to escape at the mouth of the jar. Regular bell-jars are desirable in such experiments, but in this and many other cases ordinary specie jars or candy jars will answer the purpose.

6. (page 34.) The diffusion of gases has been accounted for by the action of the repulsive force. Two gases, hydrogen and carbonic acid, for instance, are brought in contact with each other. The molecules of each of these gases are, as we have seen, a good way apart, and they tend to separate still farther, owing to the excess of the repulsive force. If there is no repulsive force acting between the molecules of carbonic acid and those of hydrogen, the molecules of hydrogen will of course work their way between and among the molecules of carbonic acid, and those of carbonic acid among those of hydrogen.

When a beam of sunlight enters a darkened room, the dust shows the molecules of the air to be in constant motion. Some have supposed that the diffusion of gases is caused by this constant agitation; just as oil and water can be intimately mixed by continued agitation.

Dalton held the first view, and Sainte-Claire Deville the second.

Others, as Miller, maintain that there is adhesion between the molecules of different gases. There seems to be no good reason for supposing that adhesion does not act between molecules in this state, since we have found it to exist in every other. Of course the excess of the repulsive force in this state and the constant agitation of their molecules would cause gases to diffuse more readily than liquids.

7. (page 41.) A small Grove's battery (269), of from four to six cells, answers very well for this and similar experiments; but Bunsen's battery (222, 269) is the most convenient, not only for these purposes, but for all the ordinary experiments in electricity.

For electrical experiments, except those with Ruhmkorff's coil (275-277), and the electric lamp (264), six Bunsen cells of medium size are amply sufficient. For working a large coil ten or twelve such cells are needed, and for the electric light from forty to fifty.

The sulphuric acid used in these batteries is diluted with from twelve to sixteen parts of water.

The water to be decomposed (47) must be slightly acidulated with sulphuric acid.

Any form of decomposing cell answers the purpose. If the one represented in Figure 19 is used, it is better without the stand than with it. In performing this experiment, remove the little bell-jars, and immerse the U-tube in acidulated water so as to fill it; then fill the bell-jars and put them into the arms while still under water.

8. (page 41.) We have found the following method most convenient for showing the composition of water by synthesis. A tubulated retort holding about a quart is first filled with oxygen by allowing a stream of that gas to pass through it for some time. The oxygen is introduced by a tube passing through a cork which accurately fits the tubulature. The neck of the retort is at the same time closed by a cork through which passes another glass tube. A rubber tube connected with this end opens under water. As the oxygen passes through the retort, the air is driven out, and since the rubber tube opens under water no air can enter. After the retort is filled with oxygen, the cork is removed from the tubulature and the glass stopper replaced. The retort is then inverted, and fastened to a clamp with its neck inclined so that water or other liquid can easily run down to its mouth. The rubber tube is then introduced into a jar of oxygen inverted over the trough. A cork which accurately fits the tubulature is then fitted over the jet of an oxyhydrogen blowpipe (130). The oxygen and hydrogen are then allowed to escape from the jet and are ignited, the flow of the gas being regulated so as to give quite a small flame. The stopper is then removed from the retort, and the blowpipe jet introduced. We then have oxygen and hydrogen burning in an atmosphere of oxygen. The neck of the retort is kept cool by wrapping a cloth round it, and allowing a stream of water to trickle down upon it. A liquid soon runs down into the neck of the retort and collects at the cork. After a time this liquid may be caught in a test-glass; and it may be shown to be water by putting a bit of potassium in a porcelain dish and dropping a little of the liquid upon it. The end of the retort is connected with the jar of oxygen by the rubber tube, in order to maintain a uniform pressure within the retort, and at the same time to prevent the air from entering it. The water produced must

of course be the product of the combination of oxygen and hydrogen.

A simpler though less satisfactory way of performing this experiment is to introduce the lighted blowpipe jet into the tubulature of the retort arranged as above, except that it has not been filled with oxygen, and its mouth not connected with a jar of oxygen. The water soon trickles down the neck of the retort, and may be collected in a test-glass. In this case a string must be wound tightly round the neck of the retort near its mouth, to prevent the water on the outside from flowing off at the end.

The management of the oxy-hydrogen blowpipe is explained in section 130, and Appendix, 23.

Oxygen is most readily prepared by mixing chlorate of potash with one fourth its weight of black oxide of manganese (205), and heating the mixture in a flask of glass, or better of copper. The explosions which sometimes occur when oxygen is prepared by this process are probably due to impurities, like carbon and compounds of antimony, in the black oxide of manganese. These impurities may be in great part burned off by heating the oxide to redness.

The oxide seems to take no part in the reaction, which is as follows : —

$$KCl\Theta_3 = KCl + 3\Theta.$$

9. (page 43.) The following is the most convenient way of introducing the gases into the tube. The tube is first filled with water; a fine glass tube is then introduced into the open end, and connected by means of a rubber tube with a hydrogen generator. The hydrogen is allowed to escape very slowly into the U-tube, which must be held inclined with the closed arm uppermost, so that the gas can bubble up into this arm. In order that no air may get into the U-tube, it must be kept under water in the vessel in which it was first filled with water; and this vessel must be large enough and deep enough for doing this conveniently. The small glass tube connected with the hydrogen generator is also plunged into this same vessel of water, and the hydrogen allowed to pass through it until we are certain that the air is all out; it is then introduced

into the open arm of the U-tube without removing it from the water. The whole can then be removed from the water, and the hydrogen allowed to pass into the closed arm. After sufficient hydrogen has passed into the tube, its quantity is marked by pasting a bit of paper on the arm of the tube. Oxygen is then introduced in the same way; but as the oxygen cannot be generated steadily enough to fill the tube, it must be first collected in some suitable vessel. The most convenient vessel is a tubulated bell-jar. The glass tube that passes into the open arm of the U-tube is connected with the bell-jar by means of a stop-cock. The stop-cock is then opened, and the bell-jar pressed down into water. This forces the gas out, and with care it can be made to pass out in a very gentle stream.

Care must be taken that all the air is driven out of the flask before the oxygen is collected, and also that the air is driven out of the glass tube before it is inserted into the U-tube. To insure this, the same precautions must be taken as above described when the hydrogen was introduced. After the oxygen is introduced, the space occupied by the mixed gases is also marked by pasting a bit of paper on the tube. The arm of the tube should never be filled more than two thirds full.

The electric spark is passed through the mixed gases by charging a Leyden jar, and connecting its outer coating with one of the platinum wires, and then bringing the knob in contact with the other. As the gases combine explosively, the thumb must be held firmly on the open end of the tube at the time of the passage of the spark; and to guard against breaking the tube, the open arm should be filled with water only within two inches of the top. The air between the water and the thumb serves as an elastic cushion. Before measuring the residual gas, the water must be brought to the same level in each arm, either by pouring in more water, or by opening the nipper-tap and allowing some of the water in that arm to escape. To examine the residual gas, the open arm of the tube must be completely filled with water and firmly closed with the thumb, and the tube so inclined as to transfer the gas to this arm. It will be difficult to introduce just half as much oxygen as hydrogen; but this is not necessary, since we need only show that just twice as much hydrogen as oxygen is always used.

10. (page 44.) For the preparation of chlorine see 13 below.

The decomposition of water by chlorine may be effected by the apparatus shown in Figure 120. "In the larger flask chlo-

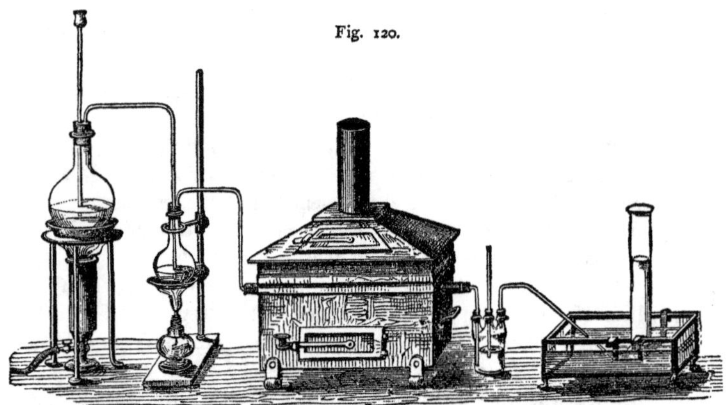

Fig. 120.

rine is evolved from hydrochloric acid by means of black oxide of manganese, and the gas escaping from this flask is caused to bubble through hot water contained in the smaller flask. The chlorine, thus saturated with steam, is then passed through a porcelain tube heated to redness in the furnace. The gas which escapes from the red-hot tube is easily recognized as a mixture of oxygen and hydrochloric acid. To separate these two gases, it is only necessary to connect with the furnace-tube a wash-bottle filled with water (or solution of soda), which absorbs and retains the hydrochloric acid, allowing the pure oxygen to pass on through the delivery-tube into the inverted cylinder." (Hofmann.)

Instead of the furnace represented in the figure, it is convenient, where gas can be had, to use one similar to that shown in Figure 32 (page 66). This consists of a row of burners, above which is a sheet-iron casing (not shown in the figure) for enclosing the tube. The casing is pierced with holes below, so that the flame may come in direct contact with the tube.

Since porcelain tubes are very liable to break, the above is on the whole an unsatisfactory experiment. A simpler method of showing that chlorine can set oxygen free from water is to fill some flasks with chlorine water, invert them in shallow vessels filled with water, set them in direct sunshine, and allow them to stand a day or so. Oxygen is set free in minute bubbles, and collects in the flasks. It can be tested in the ordinary way. The chlorine water is prepared by filling the flasks with water, and causing chlorine to pass through them until the liquid acquires the characteristic color and smell of that gas.

11. (page 50.) Precisely the same precautions to prevent any air from going into the cylinder must be taken here as in filling a jar with ammonia gas (2). Everything in this process is identical with the one there described. Of course the mouth of the jar must be closed with the glass plate when it is removed from the mercury and introduced into the water. See Figure 121.

Fig. 121.

Sodium amalgam, used in the next experiment (page 51), if it cannot be bought, may be made by moderately heating some mercury in a small beaker glass, and adding cautiously and slowly bits of sodium. The experiment requires only a two-ounce bottle two thirds full of the amalgam; and to prepare this amalgam would require only three or four bits of sodium of the size of a pea. All air should be driven from the flask before it is connected with the sodium amalgam bottle.

12. (page 52.) This apparatus can be filled with dilute muriatic acid by the method described above (7). When the bell-jar becomes filled with hydrogen, it should be removed, else the hydrogen will pass across the bend into the other jar.

13. (page 52.) Three ounces of the powdered black oxide of manganese, with half a pint of commercial muriatic acid

diluted with 3 ounces of water, will yield from 3 to 4 gallons of the gas.

14. (page 54.) The composition of muriatic acid can be shown by the method described above (9) for showing the composition of water, though less accurately, since chlorine is quite soluble. If, however, the chlorine is introduced into the U-tube last, and the experiment performed rapidly, the results are sufficiently exact for class work. We always use this method, and find it accurate enough, and on the whole much more satisfactory than the one given in the text.

15. (page 56.) We have found it more convenient to use an ordinary test-tube in place of the test-bulb. The test-tube is closed with a cork through which pass two tubes, one to connect it with the drying-bottle, the other with the jar in which the gas is collected. The first tube must pass to within a quarter of an inch of the bottom of the test-tube; the other just through the cork. Of course in all such experiments it is desirable to have the glass tubes merely pass outside of the cork far enough for attaching a rubber tube, so that the connections may be mainly of rubber; thus lessening the liability of accident by breakage.

A piece of potassium about the size of a pea is needed to decompose ammonia enough to fill with hydrogen a cylinder one inch in diameter and five inches high. The potassium must be carefully dried from naphtha by means of unsized paper before it is put into the test-tube. After it is put in, the cork is replaced, the ammonia gently boiled in the flask, and the gas allowed to run through the apparatus till the air is all driven out. This can be ascertained by holding the end of the delivery-tube under water. So long as there is air coming over, bubbles rise to the surface of the water; when the air is all out, no bubbles rise, since the ammonia is absorbed by the water. The end of the delivery-tube is then held under the mouth of the jar in which the hydrogen is to be collected, and the end of the test-tube in which the potassium has been put is cautiously heated by means of the lamp. The potassium should be heated just enough to keep it in a molten state. If the hydrogen is collected over water, the delivery-tube must be removed from the water before removing the lamp, else the water will absorb the gas in the tube, and flow back into the test-tube and break it.

16. (page 65.) As phosphorus is a very inflammable substance, and as burns from it are very slow to heal, great caution must be exercised in using it. The sticks must always be cut under water, and it must be carefully dried, by pressing it between pieces of unsized paper, before it is burnt; else it will fly about in a very disagreeable and dangerous manner. It should always be lighted by touching it with a heated wire, never by holding it in the flame of a lamp.

The above experiment may be performed as follows. A lump of chalk is cut into a cylinder some 3 inches high and 1½ inches in diameter; the top is hollowed out into a cup from ½ to ¾ of an inch deep. This cylinder is placed upright in a shallow dish partially filled with water. A piece of phosphorus about the size of a large bullet is placed in the cup, and ignited by touching it with a red-hot wire. A jar of oxygen is then quickly inverted over it. The jar should have a ground mouth, and the mouth should be closed with a glass plate, which is removed just as the jar is inverted. The best effect of the light is obtained in a glass globe; but, as the jar is liable to get broken in the experiment, we ordinarily use a good-sized specie jar. After being filled with oxygen, the jar should stand till it has drained quite dry.

17. (page 65.) Asbestos is an incombustible mineral substance, of a fibrous structure, somewhat resembling flax. When it is *platinized* it is coated with a film of platinum powder. It should be enclosed in a short tube of hard glass. The gases are mixed in a tubulated bell-jar over the water-trough, and the tube containing the platinized asbestos is connected with the stop-cock of the bell-jar. The asbestos is heated by means of a lamp, and the mixed gases forced through it by opening the stop-cock and pressing the jar down into the water.

18. (page 66.) In this experiment the jars should be as dry as possible, since the red fumes are absorbed by water, and

quickly fade out when there is water on the sides of the jars. A jar is first filled with nitrous oxide, and placed upright with its mouth covered with a glass plate (see Figure 34, page 96). A jar of half the capacity is then filled with oxygen, and placed inverted upon the glass plate covering the other jar. The plate is then removed, and the mouths of the jars brought together. The jars should be of the same diameter, and their mouths should be carefully ground, so as to prevent the fumes from escaping into the room. These red fumes are very poisonous.

19. (page 66.) A piece of an iron gas-pipe ($\frac{3}{4}$ inch is a good size) is better than a porcelain tube for this experiment, since it is less liable to break; and a retort is better for boiling the acid than a flask, as the fumes act upon the cork. The neck of the retort is run into the iron tube, and the joint made air-tight by means of putty. It is necessary to insert a large glass tube in the tube which delivers the gas from the heated tube, in order to show the red fumes.

Marsh-gas pure enough for ordinary purposes may be very easily and cheaply prepared by passing the vapor of alcohol through the red-hot iron tube of the apparatus just described. The alcohol may be boiled in the retort, and it is well to pass the gas several times through the tube. We recommend this way of obtaining the gas, as the mixing of the lime and caustic soda for the other method is a disagreeable task.

20. (page 89.) This is usually called the law of *elective affinity*. The expression is objectionable, since we know of no chemical affinity that is not elective.

21. (page 96.) Such a gas-burner may be arranged by bending a piece of glass tubing at right angles, and drawing one end to a fine opening. This is connected by a rubber tube to an ordinary gas-jet, and then the fine-drawn end is put up through one of the holes in the shelf of the pneumatic trough, and the gas is let on in a gentle stream, and lighted as it issues from the glass tube. A bell-jar of air is then inverted over it. To show that the water will rise in the jar, it is better to shut off the gas before it has ceased to burn, else the gas escaping into the jar

will take the place of that which has been removed by the burning.

22. (page 96.) This experiment is best performed by putting a piece of well-dried phosphorus in a little porcelain dish which is floating on the surface of water, igniting it by means of a heated wire, and inverting a jar of air over it. Before introducing the nitric oxide, care must be taken that the air has all been driven from the tube which delivers this gas, else it will give a red color to the gas in the bell-jar.

23. (page 100.) The gases are first mixed in a tubulated bell-jar. This can be easily done by filling the bell-jar first with water, and then passing into it one measure of oxygen and two of hydrogen. The rubber gas-bag is then connected with the bell-jar by means of a stop-cock and connecting screw, the stop-cock is opened, and the bell-jar pushed down into the water. By this means the gas is forced from the jar into the bag.

Fig. 122.

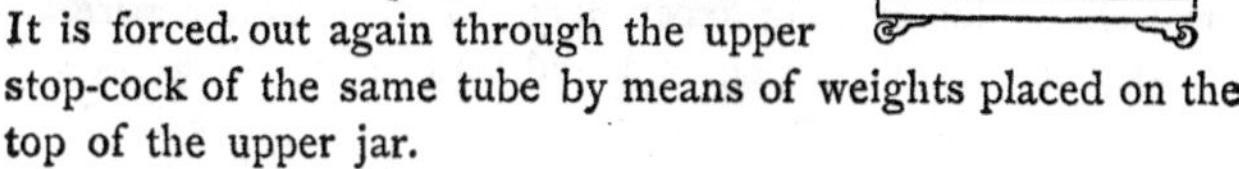

The stop-cock of the rubber bag should always be closed before the bubbles are ignited, to prevent the possibility of setting fire to the gas.

The construction of the gasometer is evident from Figure 122. The lower jar is constructed with a core, so as to require the least possible quantity of water. When the upper jar is down it is filled with water, and as it fills with gas it rises. The gas passes in through the tube seen in the centre of the core, which is connected with the stop-cock on the outside. It is forced out again through the upper stop-cock of the same tube by means of weights placed on the top of the upper jar.

In using the bell-jar, especially with hydrogen, great care must be taken that all the joints are well packed and perfectly air-tight; for when this gas is mixed with air it is very explosive. Hydrogen should never be used long after it has been put into one of the jars, since it may become mixed with air.

24. (page 102.) A specie jar had better be used for this experiment, since it is very liable to be broken by the melted product of the combustion. Water should be left in the jar to the depth of an inch or two.

It is convenient in this and the following experiment to have a disc of sheet tin or copper with a hole pierced in the centre. The end of the watch-spring is passed through the hole and then through a cork above, and thus fastened to the disc.

25. (page 102.) A deflagrating spoon may be made of brass or copper. It should be fastened to a metallic disc by means of a cork, as described above (24). The potassium or sodium must be well dried, and the oxygen must be thoroughly washed by passing it through a wash-bottle of water, and dried by dropping a few bits of fused chloride of calcium into the jar sometime before the experiment is tried. The potassium or sodium is put into the spoon, heated in the flame of the lamp, and quickly plunged into the jar of oxygen.

26. (page 116.) Of course the phosphorus must be scraped under water (16). The jar is cleansed from grease by pouring into it a little strong sulphuric acid, and turning the jar in such a way as to bring the acid in contact with every part of the surface. The jar is then thoroughly rinsed with clear water. The jar must be kept at a temperature of about 70° during the preparation of ozone.

27. (page 116.) Small pieces of diamond which are unfit for cutting can be obtained for about a dollar apiece, and answer very well for this striking experiment. Fine platinum wire is wound round the bit of diamond, and fastened to a larger wire, which is attached to a metallic disc. The diamond is then heated white hot in the flame of a lamp, and quickly plunged into a jar of oxygen.

A piece of graphite from a lead pencil answers very well for the experiment. It is fastened to a platinum wire in the same way as the diamond, heated white hot, and put into the oxygen. It burns much less brilliantly and for a shorter time, but usually produces enough carbonic acid to give a decided reaction with lime-water.

28. (page 88.) These examples have been chosen with special reference to their simplicity. It must not be supposed that all chemical reactions are equally simple.

29. (page 120.) "It is from one of the nitrates that the acid is always obtained for chemical purposes. When nitrate of potassium is heated with a powerful acid, such as the sulphuric, mutual decomposition occurs. The potassium and the hydrogen change places, forming sulphate of potassium and nitrate of hydrogen, or nitric acid: —

$$2KNO_3 + H_2SO_4 = K_2SO_4 + 2HNO_3.$$

The sulphate of potassium remains in the retort, whilst the more volatile nitric acid distils over, and may be condensed in the usual manner. The method of procuring nitric acid offers a good example of the general principle upon which acids which admit of being distilled without experiencing decomposition are obtained from their salts. In preparing nitric acid on the small scale, equal weights of nitre and oil of vitriol are placed in a glass retort and distilled. During the distillation red fumes appear in the retort, arising from a partial decomposition of the acid, and a formation of some of the lower oxides of nitrogen, whilst a yellowish corrosive liquid is condensed in the receiver; this liquid is concentrated nitric acid; it fumes strongly in the air, and emits a powerful irritating acid odor.

On the large scale, iron retorts, coated with fire-clay on the inside of the upper part, where they are exposed to the acid vapors, are employed for the distillation, and nitrate of sodium is substituted for nitrate of potassium, as it is a cheaper salt, and likewise yields nine per cent more nitric acid than nitrate of potassium." (Miller.)

30. (page 160.) The preparation of hydrochloric acid is sometimes a special branch of chemical manufacture. "It is easily prepared for use by placing in a capacious retort 3 parts of fused chloride of sodium in fragments, and introducing gradually, through a bent funnel, 3 parts of oil of vitriol. If pounded salt be used, the action of the acid is apt to be too rapid. The retort is connected with a series of Woulfe's bottles; in the first, a small quantity of water is placed to detain any impurities which might be carried over mechanically with the gas; the second bottle may contain 4 parts of water, and should be immersed in a vessel of cold water, as the condensation of the gas is attended with a great disengagement of heat. On applying a gentle heat to the retort, the acid comes over and is condensed; an easily soluble acid sulphate of sodium remains in the retort. For manufacturing purposes the decomposition is effected in iron cylinders, like those employed in the manufacture of nitric acid, and only one half the quantity of sulphuric acid prescribed above is used: —

$$2NaCl + H_2SO_4 = 2HCl + Na_2SO_4.$$

The acid in this case is in the proportion of one equivalent to each equivalent of salt, neutral sulphate of sodium remaining in the cylinder, whilst the acid is condensed in a series of salt-glazed stoneware jars, arranged as Woulfe's bottles." (Miller.)

31. (page 247.) Here, as elsewhere, we have described only one or two experiments, which serve to illustrate the principles. If the teacher has the apparatus for a larger number of experiments, he will of course make use of it at the proper point; if he has not, it is hardly worth while that the pupil should learn descriptions of experiments which he never witnesses.

A very pleasing illustration of the electric light in rarefied air is afforded by the "guinea and feather tube" used in pneumatic experiments. If the ends of the tube are connected with the poles of the inductorium (or with the electrical machine) purple flashes of auroral light mark the passage of the current through the tube when the air is exhausted. In all experiments of this kind, the room should be darkened.

Gassiot's cascade is a simple and inexpensive piece of apparatus for showing the electric light in a vacuum. It consists of a large glass goblet (uranium glass is best), the inside of which is coated nearly to the top with tinfoil. Place the vessel on the plate of the air-pump, cover it with a receiver which has a sliding rod through the top, bring the sliding rod in contact with the tinfoil coating, and connect one pole of the inductorium (or one conductor of the electrical machine) with the rod, and the other with the pump-plate. When the air is exhausted, and the current sent through the receiver, streams of blue light flow from the tinfoil over the side of the vessel to the pump-plate. A variety of beautiful effects are produced by different degrees of exhaustion, and by changing the direction of the current.

The apparatus known as the *Abbé Nollet's Globe* also furnishes very pretty displays of the electric light in rarefied air. It consists of a glass globe suspended in the upper part of a glass bell-jar, and arranged so that it can be partially filled with water; and connected with the inductorium or the electrical machine by means of a chain dipping into the water. The light in this case flows in lambent streams from the globe to the pump-plate.

A variety of pieces of apparatus for showing the electric light are made by pasting bits of tinfoil about $\frac{1}{20}$ of an inch apart on glass, oiled silk, or other non-conducting substance. Letters, outline figures, etc., may thus be formed, which appear in lines of scintillating light when the current is sent through them.

The pieces of tinfoil may be pasted in a spiral on the inside of a long glass tube, and lighted up in the same way.

The *diamond jar*, as it is sometimes called, is a Leyden jar, the coatings of which are composed of small pieces of tinfoil, separated from one another. Brilliant sparks pass between these pieces when the jar is charged or discharged.

If the knob of a common Leyden jar is connected with one pole of the inductorium, and a wire from the other pole is brought near the outer coating of the jar, bright sparks pass in most rapid succession between the pole and the jar. The electrical machine may be used instead of the inductorium in this experiment, but the effect is much less striking.

The teacher will find many other experiments in the works

on Electricity mentioned in the Preface, especially in the little book of Ferguson's. To those who have Ruhmkorff's coil, we commend a little volume by Noad, entitled "The Inductorium," (London, John Churchill and Sons, 1866,) which describes a large number of beautiful and instructive experiments with that instrument.

32. (page 228.) The zinc used for battery purposes should in all cases be amalgamated. This may be done either by immersing the zinc in mercury, or by rubbing its surface with that metal. In either case the zinc should first be cleaned with dilute sulphuric acid.

It is well to amalgamate the zinc plates of a battery every time it is used, and the best time for doing this is when the battery is taken down after being used, as the zincs then need no cleaning. For amalgamating the zincs of a large Bunsen's battery, a cylindrical vessel of soapstone, made with a core and just large enough for immersing the zincs in the mercury, will be found convenient. With such a vessel, not more than forty pounds of mercury will be needed. After immersion in the mercury the zincs should be set to drain in an iron sink, the surplus mercury being caught in a vessel below.

FRENCH WEIGHTS AND MEASURES.

LINEAR MEASURE.

Millimètre	=	0.03937	inch.	or	0.003281	ft.
Centimètre	=	0.39371	"	"	0.032809	"
Décimètre	=	3.93708	"	"	0.328090	"
Mètre	=	39.37079	"	"	3 280899	"
Décamètre	=	393.70790	"	"	32.808992	"
Hectomètre	=	3937.07900	"	"	328.089920	"
Kilomètre	=	39370.79000	"	"	3280.899200	"

SQUARE MEASURE.

Centiare, or square mètre	=	10.764299	square ft.
Are, or 100 square mètres	=	1076.429934	"
Hectare, or 10,000 square mètres,	=	107642.993418	"

CUBIC MEASURE.

Millilitre, or cubic centimètre	=	0.06103	cubic inch.
Centilitre, or 10 cubic centimètres	=	0.61027	"
Décilitre, " 100 " "	=	6.10271	"
Litre, or cubic décimètre	=	61.02705	"
Décalitre, or centistère	=	610.27052	"
Hectolitre, or décistère	=	6102.70515	"
Kilolitre, or *stère*, or cubic mètre	=	61027.05152	"

WEIGHTS.

Milligramme	=	0.01543	Troy grain.
Centigramme	=	0.15432	"
Décigramme	=	1.54323	"
Gramme	=	15.43235	"
Décagramme	=	154 32349	"
Hectogramme	=	1543.23488	"
Kilogramme	=	15432.34880	"

☞ In all these tables, 10 of each denomination make 1 of the next denomination.

QUESTIONS FOR REVIEW AND EXAMINATION.

COHESION.

1. WHAT are molecules? 2. Prove that molecules are exceedingly minute. 3. Show that they are not in actual contact. 4. What is probably true of the spaces between them? 5. Show that an attractive force acts between the molecules of a body. 6. Show that a repulsive force acts between the molecules of some bodies. 7. Show that these two molecular forces sometimes act together. 8. What is supposed to be true of them in all cases? 9. What are the three states of matter. 10. What is the relative strength of the two molecular forces in each? 11. What is cohesion? 12. What is adhesion? 13. Show that these forces act only through insensible distances. 14. From what do the properties of solids result? 15. What is meant by tenacity? 16. How can the relative tenacity of different solids be determined? 17. When is a solid said to be hard? 18. When soft? 19. When is a solid said to be elastic? 20. In what do solids differ, as regards elasticity? 21. When is a solid said to be brittle? 22. When ductile or malleable? 23. Show that solids are somewhat compressible. 24. Show that the cohesive force, when free to act, often arranges the molecules of a solid in regular forms? 25. What are these regular forms called? 26. Show that the molecules of a solid cohere more strongly on some sides than on others. 27. What are Rupert's drops? 28. Why must glass be annealed? 29. How is this done? 30. How is steel tempered? 31. What is the distinguishing characteristic of liquids? 32. What is true of the spaces between the molecules of a given body in the solid and in the liquid state? 33. Prove this. 34. Show that liquids are but slightly compressible. 35. Show that the molecules

of liquids when left to themselves tend to collect into spheres. 36. What is the distinguishing characteristic of gases? 37. Show that the molecules of a given body are farther apart in the gaseous than in the liquid state. 38. Show that gases are readily compressible.

ADHESION.

39. Show that there is adhesion between solids and solids. 40. Show that different solids adhere to each other with different degrees of strength. 41. Show that the adhesion between the solids is sometimes strong enough to overcome the cohesion of one of the solids. 42. Show that there is adhesion between solids and liquids. 43. What are the three cases of this adhesion? 44. Illustrate each. 45. Show that the surface of a solid is increased by pulverizing it. 46. How is sugar refined? 47. What promotes solution? 48. Prove this. 49. Show that different solids are not equally soluble in the same liquid. 50. Show that the same solid is not equally soluble in different liquids. 51. Illustrate capillarity. 52. What are the different cases of capillarity? 53. Show that the capillary force is a very strong force. 54. Show that it never causes a liquid to flow through a tube. 55. Explain the action of a lamp-wick. 56. Show that there is adhesion between solids and gases. 57. Show that the same gas adheres to different solids with different degrees of strength. 58. Show that the same solid adheres to different gases with different degrees of strength. 59. Show that there is adhesion between liquids and liquids. 60. Illustrate the diffusion of liquids. 61. Illustrate the osmose of liquids. 62. How may some cases of osmose be explained? 63. Show that there is adhesion between liquids and gases. 64. What helps a liquid to absorb a gas? 65. Explain this. 66. Explain how by the agency of heat a gas can be separated from a liquid which has absorbed it. 67. Illustrate the diffusion of gases. 68. Illustrate the osmose of gases.

CHEMICAL AFFINITY.

69. What takes place when potassium or sodium is thrown upon water? 70. What gas is thus set free from the water? 71. How can this gas be recognized? 72. Prove that there is oxygen in water. 73. Prove that hydrogen and oxygen are the only substances in water. 74. What are compound substances? What are elements? 75. What is Affinity? 76. How does it differ from Adhesion? 77. Show that there are two measures of hydrogen in water to one of oxygen. 78. Show that affinity is stronger between some substances than others. 79. Show that a molecule of water can be divided into three, and only three parts. 80. What is an atom? 81. What are the relative weights of the smallêst parts into which hydrogen, oxygen, sodium, potassium, and chlorine can be divided? 82. How do we know that an atom of oxygen weighs sixteen times as much as an atom of hydrogen? 83. What is the threefold divisibility of matter? 84. What is taken as the symbol for an element? 85. What does it indicate? 86. How is the symbol of a compound formed? 87. What does it indicate? 88. Show that ordinary muriatic acid contains a gas. 89. Show that this gas is readily soluble in water. 90. Prove that this gas contains hydrogen. 91. Prove that muriatic acid contains chlorine. 92. Prove that muratic acid contains only hydrogen and chlorine. 93. How can chlorine be recognized? 94. Show what proportions of hydrogen and of chlorine are contained in muriatic acid. 95. Show that a molecule of muriatic acid contains only two atoms. 96. Show that an atom of chlorine weighs 35.5 times as much as an atom of hydrogen. 97. Prove that ammonia gas contains hydrogen. 98. Prove that it contains nitrogen. 99. How is nitrogen distinguished from the other gases that we have become acquainted with? 100. Prove that ammonia contains only hydrogen and nitrogen. 101. Show the proportions of hydrogen and nitrogen in ammonia. 102. Show that a molecule of ammonia contains 4 atoms, of which 3 are hydrogen. 103. Show that the atomic weight of nitrogen is 14. 104. How may marsh-gas be prepared? 105. Prove that it contains carbon. 106. Prove that it contains hydrogen. 107. Prove that it

contains only those two elements. 108. What proportions of hydrogen and carbon in marsh-gas? 109. Show that a molecule of marsh-gas contains 4 atoms of hydrogen and one of carbon. 110. How do the different elements compare with one another in their atom-fixing power? 111. What is a monatomic element? 112. A diatomic element? 113. A triatomic element? 114. A tetratomic element? 115. Name the monatomic elements. 116. Name the diatomic elements. 117. The triatomic elements. 118. The tetratomic elements. 119. Which of the elements have barred symbols? 120. Show that nitric oxide contains oxygen. 121. Show that it contains nitrogen. 122. Show also that it contains only oxygen and nitrogen. 123. Show that nitrous acid contains oxygen. 124. That it contains nitrogen. 125. Show that nitric acid contains oxygen. 126. That it contains nitrogen. 127. How do nitric oxide, nitrous acid, and nitric acid differ from one another? 128. How do hydrogen and nitrogen differ in the compounds which they form? 129. Show that the compounds of oxygen are sometimes bases, sometimes acids, and sometimes neutrals. 130. How are acids which contain oxygen named? 131. How are the bases and neutrals which contain oxygen named? 132. How do oxyacids and hydracids differ? 133. Give the names and symbols of the leading hydracids. 134. Of the leading oxyacids. 135. Of the bases of each of the four groups. 136. Of the sulphides of the metals of each of the four groups. 137. Of the chlorides of the same. 138. In what form do the oxyacids ordinarily exist? 139. Explain the two methods of writing the symbols of these compounds. 140. How do the bases usually exist? 141. Show how the symbols of these compounds of the bases are written. 142. How is the symbol for hydrate of ammonia usually written? 143. Explain this symbol. 144. What are compound radicals? 145. What compound radicals are mentioned? 146. Can they always exist in a free state? 147. What are the name and symbol of the substance formed out of nitric acid and soda? 148. What are substances which have an analogous composition called? 149. Explain the two methods of writing the symbols of these compounds. 150. Explain the action of the hydrates of the acids upon the hydrates of the bases. 151. Illustrate this action by showing the action of the hydrates of sulphuric and nitric acid on the hydrate of a base in each of the four groups. 152. Define normal, basic, and acid salts. 153. Define monobasic, dibasic, and tribasic acids. 154. Explain the different ways of forming ternary salts. 155. Give an illustration of each of these methods. 156. Show what is meant by substitution, by single decomposition, and by double decomposition. 157. Show the action of hydracids on a base of each of the four groups. 158. What name is given to the binary compounds thus formed? 159. Explain the different methods of forming binary salts, illustrating each by an equation. 160. State the law of double decomposition. 161. Show that the products of the combustion of coal-gas, or of a candle, are carbonic acid and water. 162. Show that the coal-gas in burning removes something from the air. 163. Show that it removes oxygen from the air. 164. Show that the burning of coal-gas consists in the combination of the free oxygen of the air with the hydrogen and carbon of the gas. 165. Show that all ordinary combustion consists in the combination of the free oxygen of the air with the burning substance. 166. Show why a draft is necessary in stoves and furnaces. 167. Show that there is no real distinction between combustibles and supporters of combustion. 168. Show what fraction of the air is free oxygen. 169. Show that oxygen must be heated before it will combine with ordinary combustibles. 170. Illustrate the passive condition of oxygen at ordinary temperatures, and the energy with which it enters into combination when it is once roused to activity by means of heat. 171. Explain how combustion is self-sustaining. 172. What is meant by the point of ignition? 173. What is true of the point of ignition of different substances? 174. Show that the products of combustion are not always gaseous. 175. Show that the burning of metals in the oxy-hydrogen flame consists in their combination with oxygen. 176. Show that some metals will burn in the air. 177. Show that oxygen is not the only supporter of combustion. 178. Show that the materials of the earth's crust are chiefly chemical compounds. 179. Show that the materials of the earth's crust are results of combustion. 180. Show that free oxygen is not necessary to combustion. 181. Explain the reduction of metallic ores by means of carbon. 182. What is the difference between slow and rapid combustion? 183. What is the slow combustion of metals called? 184. Show that the slow combustion of metals is attended with development of heat. 185. Show that decay is slow combustion. 186. Show that the decay of vegetable substances develops heat. 187. Show that animals in breathing remove oxygen from the air. 188. Show that carbonic acid is not the only product of respiration. 189. In what respects do the products of the rapid combustion of vegetable substances, and of their decay, and of respiration differ? 190. Show that vegetable substances contain carbon, hydrogen, oxygen, and nitrogen. 191. Which are the chief elements found in vegetable substances? 192. How do animal substances differ from vegetable? 193. What are the three conditions of oxygen? 194. Explain how the decay of vegetable substances begins. 195. Show how ozone may be prepared, and how it differs from ordinary oxygen. 196. Show that other elements exist in conditions quite unlike one another. 197. What are these states called? 198. Explain how it is that the constituents of the air are constant. 199. Explain the growth of plants. 200. Show what fraction of the air is nitrogen. 201. Explain how nitrogen can exist in the air in a free state. 202. Which are called the atmospheric elements? 203. Why are they sometimes called organic elements? 204. What takes place when vegetable substances are heated in close vessels? 205. Explain the properties of charcoal. 206. What are the chief products of the destructive distillation of wood? 207. Describe the chief ingredients of wood tar. 208. Explain the formation of mineral coal. 209. What are the chief differences between anthracite and bituminous coal. 210. How has the probable formation of each of these varieties of coal been illustrated? 211. How do the products of the distillation of bituminous coal differ from those of the distillation of wood? 212. What three classes of substances are found in coal tar? 213. What is the chief ingredient of the first? 214. Of the second class? 215. Of the third class? 216. What is the symbol for benzole? 217. How is aniline prepared from benzole? 218. Of what are toluole and cumole the chief ingredients? 219. What is the probable origin of petroleum? 220. State how it is obtained.

221. Give some account of the early history of gas-lighting. 222. Describe the manufacture of illuminating-gas. 223. Explain the gas-meter. 224. What substances burn with flame? 225. Show that all solids which appear to burn with flame can be converted into inflammable gases by heat. 226. Show that the presence of a solid and intense heat are necessary to illumination. 227. Show how these two conditions are fulfilled in the burning of coal-gas. 228. Describe Bunsen's lamp, and explain its action. 229. What is the best shape for a gas-flame, and why? 230. Describe the Bude light, and account for its great brilliancy. 231. Show that the burning candle is a miniature gas-factory. 232. Describe the three parts of the candle flame. 233. Describe a mouth blowpipe. 234. Of what parts does the blowpipe flame consist? 235. Why is one part called the oxidizing, and another the reducing flame? 236. What compounds are commonly known as the alkalies? 237. Which are called the fixed alkalies, and which is the volatile alkali? 238. Show that soaps, oils, and fats are salts. 239. What is the difference between hard and soft soaps? 240. How is potash prepared? 241. How was soda-ash formerly prepared? 242. From what is it now made? 243. What are the two processes in its manufacture? 244. Describe and explain each. 245. How was sulphuric acid formerly made? 246. Describe and explain its present manufacture. 247. Show that hydrochloric acid is an incidental product of the salt-cake process. 248. How is it now saved? 249. What is the chief use of the chlorine which the acid contains? 250. Explain this use. 251. Explain the preparation and use of bleaching-powder. 252. What are the chief products of the soda-ash manufacture, and what the chief materials used? 253. What are the sources of common salt, and how is it obtained from each of these sources? 254. What are the sources of sulphur? 255. How is crude sulphur purified? 256. What are the usual forms of sulphur, and how is each prepared? 257. How is ammonia usually obtained? 258. Describe the preparation of aqua ammonia. 259. How is sal ammoniac usually obtained?

ELECTRICITY.

260. What is the distinguishing characteristic of a magnet? Illustrate. 261. Show that the magnetic force resides chiefly at the ends of a magnet. 262. Show that the forces at the opposite ends of a magnet act in opposite directions. 263. What is meant by the north and south poles of a magnet? Illustrate. 264. What is the effect on a piece of soft iron of bringing it into contact with a magnet? On a piece of steel? 265. What is a loadstone? 266. Whence does a magnet derive its name? 267. Of what are ordinary magnets made? 268. What are their usual forms? 269. State what takes place when a small horizontal or dipping needle is moved alongside a bar magnet. 270. Show that the earth acts like a large magnet. 271. What is the effect of like and unlike poles of magnets on each other? 272. What name do the French give to the north pole of the magnet? Why? 273. How are magnets made? 274. Describe a cell of Bunsen's battery. 275. Prove that a force exists in the wire which joins the carbon and the zinc of this cell. 276. What is this force called? 277. Why is it often called a current? 278. When we speak of the direction of the current, from which pole of the cell do we always consider it as starting? 279. What do we mean by the poles of the cell? 280. What is an electric battery? 281. How many forms of the battery, and how do they differ? 282. What is meant by the poles of the battery? 283. What by the circuit? 284. Prove that the current meets with resistance in passing through fine wire. 285. How are substances divided as regards electricity? 286. What are the two qualities of electricity? 287. Define each. 288. Does the electric force always possess these qualities in the same proportions? 289. Illustrate this last. 290. What names are given to the two forms of the battery? Why? 291. What is a rheotome? a rheotrope? Describe each. 292. State the effect of the current upon the magnetic needle when it flows over, under, and around it. 293. Show that the pole of a magnet tends to rotate round a conductor through which a current is passing, and *vice versa*. 294. What is a rheoscope? 295. When the current deflects the needle, what

resistance does it have to overcome? 296. Describe the astatic needle. 297. Is the simple or the astatic needle more sensitive to the action of the current? Why? 298. Explain in what ways the needle can be made more sensitive to the action of the current. 299. Describe the astatic galvanometer. 300. Describe the tangent galvanometer. 301. What is a telegraph? 302. By what force are telegraphs worked? 303. What are the four parts essential to every telegraph? 304. Describe the needle telegraph. 305. When was this telegraph invented, and where is it used? 306. Describe the rheostat, and show how the resistance of a conductor can be determined by it. 307. In what way is the resistance of a conductor affected by its length and its diameter? 308. Prove that the electric current has power to render a piece of soft iron temporarily magnetic. 309. What is a helix? 310. Which end of the soft iron is made a north pole, depends upon what? 311. What is an electro-magnet? 312. What is the usual form of electro-magnets? 313. What is the strength of electro-magnets compared with that of constant magnets? 314. Show that the wire through which a current passes is magnetic. 315. Describe Page's rotating machine. 316. Upon what properties of constant and electro-magnets does its action depend? 317. Describe Froment's electro-magnetic engine. 318. Describe Foucault's self-acting rheotome. 319. Explain its action. 320. Why have not electric engines come into general use? 321. Explain how the electric force may be employed to regulate the motion of clocks. 322. Describe the sending and receiving instrument of Morse's telegraph. 323. By what other name is this telegraph known, and why? 324. State the use of the local battery, and explain the relay magnet. 325. Explain how several stations are connected in one circuit. 326. Describe the sending and receiving instruments of the Combination Printing Telegraph. 327. Explain how by means of the electric fire-alarm the location of the fire is telegraphed to the central station. 328. Explain how this location is telegraphed to the public. 329. Prove that the electric current can decompose chemical compounds. 330. Define electrolysis, electrolyte, electrode, anode, cathode, ion, anion, and cation. 331. What is the anion and what the cation when water is decomposed? 332. Show by an equation the decomposition of water by the electric current. 333. Show by an equation and explain what takes place when sulphate of copper is decomposed by the current, when the anode is of platinum. 334. The same, when the anode is of copper. 335. The same, when cyanide of silver is decomposed, and the anode is of platinum. 336. The same, when the anode is of silver. 337. The same, when cyanide of gold is decomposed, and the anode is of platinum. 338. The same, when the anode is of gold. 339. What is electrotyping? 340. Explain the process. 341. What is electro-plating. 342. Explain the process. 343. What is electro-gilding? 344. Explain the process. 345. Describe the sending and receiving instruments of Bain's electro-chemical telegraph. 346. Describe the voltameter, and explain its use. 347. Prove that the current develops heat, when it meets with resistance in passing through a wire. 348. To what is the heat developed proportional? 349. Explain how the electric force may be used in blasting. 350. Explain Duboscq's electric lamp. 351. What is true of the intensity of the light and heat obtained from this lamp? 352. Explain the electro-thermal telegraph. 353. When a piece of amalgamated zinc and a piece of platinum are joined at one end by a wire, and their free ends are plunged into dilute sulphuric acid, what is obtained? 354. What do two metals arranged in this way constitute? 355. Prove that in such cases the current is also passing through the liquid. 356. What is the direction of the current inside the liquid as compared with that outside? 357. When the amalgamated zinc and the platinum are joined outside the cup what action is going on in the inside? 358. Explain this action in full by means of an equation. 359. What are the two theories with reference to the origin of the electric force of the cell? 360. Describe an experiment which proves that the current can be obtained without contact of dissimilar metals. 361. Prove that without chemical action there is no current; and that the current starts, the moment chemical action begins. 362. Upon what does the direction of the current always depend? 363. Which theory of the origin of the current in the cell is now generally received? 364. How is the fact that there is no apparent chemical action in the cell before the current starts, while there is such action afterwards, explained by each theory?

365. Why is no current obtained when only one metal and one liquid are used? 366. Why do not copper and zinc constitute so good a voltaic pair as platinum and zinc? 367. Describe Grove's gas battery. 368. What general fact does it illustrate? 369. About which plate in the cell do the ions tend to collect? 370. Why do they not collect about the other? 371. What are the objections to their collecting about either plate? 372. What is the active plate in all ordinary batteries? 373. What is the ordinary liquid used to act upon the plates? 374. What is the ion that tends to collect about one of the plates? 375. How is the collection of this ion about the plate prevented in Bunsen's battery? 376. How does Grove's battery differ from Bunsen's? 377. To whom is the principle of this battery due? 378. Describe a cell of Daniell's battery. 379. Explain the use of the porous cup. 380. By what names is the electricity developed by chemical action known? 381. Why has it been so named? 382. How many cases of the development of electricity by magnetism, and in what do they differ? 383. Illustrate the development of electricity in each case. 384. Upon what does the direction of the current developed in each case depend? 385. Which case is the most important? 386. In every magneto-electric machine, there must be a means of accomplishing what? 387. How many ways of doing this, and what are they? 388. Describe a magneto-electric machine of each kind, and explain its action. 389. Describe Ruhmkorff's induction coil. 390. What experiment shows that the electricity obtained from this coil has much greater intensity than that obtained from the battery? 391. When a copper disc is rotated under a magnetic needle, what takes place? 392. Explain this action of the needle. 393. What takes place when a copper and iron wire are joined, and the junction heated or cooled? 394. How is this shown? 395. What do two metals joined in this way constitute? 396. What name is given to electricity developed by heat? 397. What is a thermo-electric battery? 398. What are the best metals for developing electricity by means of heat? 399. What is the chief use of the thermo-electric pile? 400. Prove that no current can be developed by heat, so long as only one metal is used, and this metal is homogeneous throughout its whole length. 401. Show that it is

possible to develop a current by heat, by using only one metal. 402. These and similar experiments lead to what general conclusions? 403. Describe the electrical machine. 404. Prove that electricity can be developed by friction. 405. What is true of the quantity and of the intensity of frictional electricity? 406. Prove this. 407. How does frictional electricity compare with voltaic electricity? 408. How does it compare with magneto-electricity? 409. Why cannot a galvanometer be used in examining frictional electricity? 410. Compare the electrical machine with the cell of a battery. 411. What is an electroscope? 412. Describe two forms of the instrument, and the action of each. 413. Prove that two kinds of electricity are developed by the machine, and that these two kinds of electricity are forces acting in opposite directions. 414. What names are given to these two kinds of electric force? 415. Give other illustrations of electricity developed by friction. 416. Prove that electricity can be developed upon conductors by friction, as well as upon non-conductors. 417. Can one kind of electricity be developed without at the same time developing the other? 418. In what case is electricity developed by rubbing two bodies together? 419. Describe in full the case of three insulated conductors arranged in a row between two balls, one of which is charged with positive and the other with negative electricity. 420. When is a body said to be polarized? 421. When to be charged? 422. When neutral? 423. What is meant by discharge? 424. How must a polarized insulated conductor be situated to become charged? 425. How to become charged with the same force as that on the polarizing body? 426. How to become charged with the opposite force? 427. Show that polarization of an insulated conductor depends upon the non-conducting medium which separates it from the charged body. 428. Give a full account of Faraday's theory of induction. 429. Show that a light insulated conductor which is free to move between two polarized conductors is thrown into commotion on discharging its forces after being polarized. 430. Show that molecules of conductors and non-conductors are thrown into commotion when discharging their forces. 431. What is disruptive discharge? 432. What is conductive discharge? 433. Show that non-conductors so called are only poor conductors. 434. Explain the electric current by Faraday's theory of induction. 435. What is electrolytic discharge? 436. Explain it according to Faraday's theory. 437. On what part of a solid conductor is the charge always found? 438. Explain this fact on the theory of induction. 439. How is the charge distributed over the surface? 440. Show that polarization rises highest in the direction of least resistance. 441. Explain the fact that the charge accumulates on the ends of oblong conductors. 442. Show that the charge which a body can receive depends upon the facilities it offers for polarization. 443. Describe the Leyden jar. 444. Explain how it becomes charged, and its condition after it is charged. 445. Explain how it is discharged. 446. Describe the Leyden battery. 447. Explain the action of points on charged conductors. 448. Explain the action of the electric wheel. 449. Who first proved the identity of lightning and ordinary frictional electricity? 450. Explain the electric action of clouds. 451. What is chain lightning? 452. What is sheet lightning? 453. Explain the use of lightning-rods. 454. What is the aurora? 455. What is known of the way in which atmospheric electricity is developed?

INDEX.

C.

D.

E.

F.

G.

H.

I.

Cambridge: Electrotyped and Printed by Welch, Bigelow, & Co.

www.ingramcontent.com/pod-product-compliance
Lightning Source LLC
LaVergne TN
LVHW020214110826
845151LV00003B/711

* 9 7 8 1 4 2 5 5 3 3 7 6 2 *